TEN QUESTIONS
ABOUT HUMAN ERROR

A New View of Human Factors
and System Safety

Human Factors in Transportation
A Series of Volumes Edited by
Barry A. Kantowitz

TEN QUESTIONS
ABOUT HUMAN ERROR

A New View of Human Factors
and System Safety

Sidney W. A. Dekker

Lund University

CRC Press
Taylor & Francis Group
Boca Raton   London   New York

CRC Press is an imprint of the
Taylor & Francis Group, an informa business

CRC Press
Taylor & Francis Group
6000 Broken Sound Parkway NW, Suite 300
Boca Raton, FL 33487-2742
270 Madison Avenue
New York, NY 10016

Taylor & Francis Group
2 Park Square
Milton Park, Abingdon
Oxon OX14 4RN

© 2005 by Taylor & Francis Group, LLC
CRC Press is an imprint of Taylor & Francis Group, an Informa business
Originally published by Lawrence Erlbaum Associates

No claim to original U.S. Government works
Printed in the United States of America on acid-free paper
10 9 8 7 6 5 4 3
ISBN 13: 978-0-8058-4745-1 (Softcover) ISBN 13: 978-0-8058-4744-4 (Hardcover)

Library of Congress Cataloging-in-Publication Data

Catalog record is available from the Library of Congress

Visit the Taylor & Francis Web site at
http://www.taylorandfrancis.com

and the CRC Press Web site at
http://www.crcpress.com

Contents

v

Acknowledgments

Just like errors, ideas come from somewhere. The ideas in this book were developed over a period of years in which discussions with the following people were particularly constructive: David Woods, Erik Hollnagel, Nancy Leveson, James Nyce, John Flach, Gary Klein, Diane Vaughan, and Charles Billings. Jens Rasmussen has always been ahead of the game in certain ways: Some of the questions about human error were already taken up by him in decades past. Erik Hollnagel was instrumental in helping shape the ideas in chapter 6, and Jim Nyce has had a significant influence on chapter 9.

I also want to thank my students, particularly Arthur Dijkstra and Margareta Lützhöft, for their comments on earlier drafts and their useful suggestions. Margareta deserves special gratitude for her help in decoding the case study in chapter 5, and Arthur for his ability to signal "Cartesian anxiety" where I did not recognize it.

A special thanks to series editor Barry Kantowitz and editor Bill Webber for their confidence in the project. The work for this book was supported by a grant from the Swedish Flight Safety Directorate.

Preface

Transportation human factors has always been concerned with human error. In fact, as a field of scientific inquiry, it owes its inception to investigations of pilot error and researchers' subsequent dissatisfaction with the label. In 1947, Paul Fitts and Richard Jones, building on pioneering work by people like Alphonse Chapanis, demonstrated how features of World War II airplane cockpits systematically influenced the way in which pilots made errors. For example, pilots confused the flap and landing-gear handles because these often looked and felt the same and were located next to one another (identical toggle switches or nearly identical levers). In the typical incident, a pilot would raise the landing gear instead of the flaps after landing—with predictable consequences for propellers, engines, and airframe. As an immediate wartime fix, a rubber wheel was affixed to the landing-gear control, and a small wedge-shaped end to the flap control. This basically solved the problem, and the design fix eventually became a certification requirement.

Pilots would also mix up throttle, mixture, and propeller controls because their locations kept changing across different cockpits. Such errors were not surprising, random degradations of human performance. Rather, they were actions and assessments that made sense once researchers understood features of the world in which people worked, once they had analyzed the situation surrounding the operator. Human errors are systematically connected to features of people's tools and tasks. It may be difficult to predict when or how often errors will occur (though human reliability techniques have certainly tried). With a critical examination of the system in which people work, however, it is not that difficult to anticipate where errors will occur. Human factors has worked off this premise ever since: The notion of designing error-tolerant and error-resistant systems is founded on it.

Human factors was preceded by a mental Ice Age of behaviorism, in which any study of mind was seen as illegitimate and unscientific. Behaviorism itself had been a psychology of protest, coined in sharp contrast against Wundtian experimental introspection that in turn preceded it. If behaviorism was a psychology of protest, then human factors was a psychology of pragmatics. The Second World War brought such a furious pace of technological development that behaviorism was caught short-handed. Practical problems in operator vigilance and decision making emerged that were altogether immune against Watson's behaviorist repertoire of motivational exhortations. Up to that point, psychology had largely assumed that the world was fixed, and that humans had to adapt to its demands through selection and training. Human factors showed that the world was not fixed: Changes in the environment could easily lead to performance increments not achievable through behaviorist interventions. In behaviorism, performance had to be shaped after features of the world. In human factors, features of the world were shaped after the limits and capabilities of performance.

As a psychology of pragmatics, human factors adopted the Cartesian–Newtonian view of science and scientific method (just as both Wundt and Watson had done). Descartes and Newton were both dominant players in the 17th-century scientific revolution. This wholesale transformation in thinking installed a belief in the absolute certainty of scientific knowledge, especially in Western culture. The aim of science was to achieve control by deriving general, and ideally mathematical, laws of nature (as we try to do for human and system performance). A heritage of this can still be seen in human factors, particularly in the predominance of experiments, the nomothetic rather than ideographic inclination of its research, and a strong faith in the realism of observed facts. It can also be recognized in the reductive strategies human factors and system safety rely on to deal with complexity. Cartesian–Newtonian problem solving is analytic. It consists of breaking up thoughts and problems into pieces and in arranging these in some logical order. Phenomena need to be decomposed into more basic parts, and the whole can be explained exhaustively by reference to its constituent components and their interactions. In human factors and system safety, *mind* is understood as a box-like construction with a mechanistic trade in internal representations; *work* is broken into procedural steps through hierarchical task anlayses; *organizations* are not organic or dynamic but consist of static layers and compartments and linkages; and *safety* is a structural property that can be understood in terms of its lower order mechanisms (reporting systems, error rates and audits, safety management function in the organizational chart, and quality systems).

These views are with us today. They dominate thinking in human factors and system safety. The problem is that linear extensions of these same notions cannot carry us into the future. The once pragmatic ideas of human factors and system safety are falling behind the practical problems that have started to emerge from today's world. We may be in for a repetition of the shifts that came with the technological developments of World War II, where behaviorism was shown to fall short. This time it may be the turn of human factors and system safety. Contemporary developments, however, are not just technical. They are sociotechnical: Understanding what makes systems safe or brittle requires more than knowledge of the human–machine interface. As David Meister recently pointed out (and he has been around for a while), human factors has not made much progress since 1950. "We have had 50 years of research," he wonders rhetorically, "but how much more do we know than we did at the beginning?" (Meister, 2003, p. 5). It is not that approaches taken by human factors and system safety are no longer useful, but their usefulness can only really be appreciated when we see their limits. This book is but one installment in a larger transformation that has begun to identify both deep-rooted constraints and new leverage points in our views of human factors and system safety. The 10 questions about human error are not just questions about human error as a phenomenon, if they are that at all (and if human error is something in and of itself in the first place). They are actually questions about human factors and system safety as disciplines, and where they stand today. In asking these questions about error, and in sketching the answers to them, this book attempts to show where our current thinking is limited; where our vocabulary, our models, and our ideas are constraining progress. In every chapter, the book tries to provide directions for new ideas and models that could perhaps better cope with the complexity of problems facing us now.

One of those problems is that apparently safe systems can drift into failure. Drift toward safety boundaries occurs under pressures of scarcity and competition. It is linked to the opacity of large, complex sociotechnical systems and the patterns of information on which insiders base their decisions and trade-offs. Drift into failure is associated with normal adaptive organizational processes. Organizational failures in safe systems are not preceded by failures; by the breaking or lack of quality of single components. Instead, organizational failure in safe systems is preceded by normal work, by normal people doing normal work in seemingly normal organizations. This appears to severely challenge the definition of an incident, and may undermine the value of incident reporting as a tool for learning beyond a certain safety level. The border between normal work and incident is clearly elastic and subject to incremental revision. With every little step away from previous norms, past success can be taken as a guarantee of future safety.

Incrementalism notches the entire system closer to the edge of breakdown, but without compelling empirical indications that it is headed that way.

Current human factors and system safety models cannot deal with drift into failure. They require failures as a prerequisite for failures. They are still oriented toward finding failures (e.g., human errors, holes in layers of defense, latent problems, organizational deficiencies, and resident pathogens), and rely on externally dictated standards of work and structure, rather than taking insider accounts (of what is a failure vs. normal work) as canonical. Processes of sense making, of the creation of local rationality by those who actually make the thousands of little and larger trade-offs that ferry a system along its drifting course, lie outside today's human factors lexicon. Current models typically view organizations as Newtonian-Cartesian machines with components and linkages between them. Mishaps get modeled as a sequence of events (actions and reactions) between a trigger and an outcome. Such models can say nothing about the build-up of latent failures, about the gradual, incremental loosening or loss of control. The processes of erosion of constraints, of attrition of safety, of drift toward margins, cannot be captured because structuralist approaches are static metaphors for resulting forms, not dynamic models oriented toward processes of formation.

Newton and Descartes, with their particular take on natural science, have a firm grip on human factors and systems safety in other areas too. The information-processing paradigm, for example, so useful in explaining early information-transfer problems in World War II radar and radio operators, all but colonized human factors research. It is still a dominant force, buttressed by the Spartan laboratory experiments that seem to confirm its utility and validity. The paradigm has mechanized mind, chunked it up into separate components (e.g., iconic memory, short-term memory, long-term memory) with linkages in between. Newton would have loved the mechanics of it. Descartes would have liked it too: A clear separation between mind and world solved (or circumvented, rather) a lot of problems associated with the transactions between the two. A mechanistic model such as information processing of course holds special appeal for engineering and other consumers of human factors research results. Pragmatics dictate bridging the gap between practitioner and science, and having a cognitive model that is a simile of a technical device familiar to applied people is one powerful way to do just that. But there is no empirical reason to restrict our understanding of attitudes, memories, or heuristics as mentally encoded dispositions, as some contents of consciousness with certain expiry dates. In fact, such a model severely restricts our ability to understand how people use talk and action to construct perceptual and social order; how, through discourse and action, people create the environments that in turn determine further action and possible assessments, and that constrain what will subsequently be seen as acceptable discourse or rational decisions. We cannot begin to understand drift into failure without understanding how groups of people, through assessment and action, assemble versions of the world in which they assess and act.

Information processing fits within a larger, dominant metatheoretical perspective that takes the individual as its central focus (Heft, 2001). This view, too, is a heritage of the Scientific Revolution, which increasingly popularized the humanistic idea of a "self-contained individual." For most of psychology this has meant that all processes worth studying take place within the boundaries of the body (or mind), something epitomized by the mentalist focus of information processing. In their inability to meaningfully address drift into failure, which intertwines technical, social, institutional, and individual factors, human factors and system safety are currently paying for their theoretical exclusion of transactional and social processes between individuals and world. The componentialism and fragmentation of human factors research is still an obstacle to making progress in this respect. An enlargement of the unit of analysis (as done in the ideas of cognitive systems engineering and distributed cognition) and a call to make action central in understanding assesments and thought have been ways to catch up with new practical developments for which human factors and system safety were not prepared.

The individualist emphasis of Protestantism and Enlightenment also reverberates in ideas about control and culpability. Should we hold people accountable for their mistakes? Sociotechnical systems have grown in complexity and size, moving some to say that there is no point in expecting or demanding individual insiders (engineers, managers, operators) to live up to some reflective moral ideal. Pressures of scarcity and competition insidiously get converted into organizational and individual mandates, which in turn severely constrain the decision options and rationality (and thus autonomy) of every actor on the inside. Yet lone antiheroes continue to have lead roles in our stories of failure. Individualism is still crucial to self-identity in modernity. The idea that it takes teamwork, or an entire organization, or an entire industry to break a system (as illustrated by cases of drift into failure) is too unconventional relative to our inherited cultural preconceptions.

Even before we get to complex issues of action and responsibility, we can recognize the prominence of Newtonian–Cartesian deconstruction and componentialism in much human factors research. For example, empiricist notions of a perception of elements that gradually get converted into meaning through stages of mental processing are legitimate theoretical notions today. Empiricism was once a force in the history of psychology. Yet buoyed by the information-processing paradigm, its central tenets have made a comeback in, for example, theories of situation awareness. In adopting such a folk model from an applied community and subjecting it to putative scientific scrutiny, human factors of course meets its pragmatist ideal. Folk models fold neatly into the concerns of human factors as an applied discipline. Few theories can close the gap between researcher and practitioner better than those that apply and dissect practitioner vernacular for scientific study. But folk models come with an epistemological price tag. Research that claims to investigate a phenomenon (say, shared situation awareness, or complacency), but that does not define that phenomenon (because, as a folk model, everybody is assumed to know what it means), cannot make falsifiable contact with empirical reality. This leaves such human factors research without the major mechanism for scientific quality control since Karl Popper.

Connected to information processing and the experimentalist approach to many human factors problems is a quantitativist bias, first championed in psychology by Wilhelm Wundt in his Leipzig laboratory. Although Wundt quickly had to admit that a chronometry of mind was too bold a research goal, experimental human factors research projects can still reflect pale versions of his ambition. Counting, measuring, categorizing, and statistically analyzing are chief tools of the trade, whereas qualitative inquiries are often dismissed as subjectivist and unscientific. Human factors has a realist orientation, thinking that empirical facts are stable, objective aspects of reality that exist independent of the observer or his or her theory. Human errors are among those facts that researchers think they can see out there, in some objective reality. But the facts researchers see would not exist without them or their method or their theory. None of this makes the facts generated through experiments less real to those who observe them, or publish them, or read about them. Heeding Thomas Kuhn (1962), however, this reality should be seen for what it is: an implicitly negotiated settlement among like-minded researchers, rather than a common denominator accessible to all.

There is no final arbiter here. It is possible that a componential, experimentalist approach could enjoy an epistemological privilege. But that also means there is no automatic imperative for the experimental approach to uniquely stand for legitimate research, as it sometimes seems to do in mainstream human factors. Ways of getting access to empirical reality are infinitely negotiable, and their acceptability is a function of how well they conform to the worldview of those to whom the researcher makes his appeal. The persistent quantitativist supremacy (particularly in North American human factors) seems saddled with this type of consensus authority (it must be good because everybody is doing it). Such methodological hysteresis could have more to do with primeval fears of being branded "unscientific" (the fears shared by Wundt and Watson) than with a steady return of significant knowledge increments generated by the research.

Technological change gave rise to human factors and system safety thinking. The practical demands posed by technological changes endowed human factors and system safety with the pragmatic spririt they have to this day. But pragmatic is no longer pragmatic if it does not match the demands created by what is happening around us now. The pace of sociotechnological change is not likely to slow down any time soon. If we think that World War II generated a lot of interesting changes, giving birth to human factors as a discipline, then we may be living in even more exciting times today. If we in human factors and system safety keep doing what we have been doing, simply because it worked for us in the past, we may become one of those systems that drift into failure. Pragmatics requires that we too adapt to better cope with the complexity of the world facing us now. Our past successes are no guarantee of continued future achievement.

Series Foreword

Barry H. Kantowitz
Battelle Human Factors Transportation Center

The domain of transportation is important for both practical and theoretical reasons. All of us are users of transportation systems as operators, passengers, and consumers. From a scientific viewpoint, the transportation domain offers an opportunity to create and test sophisticated models of human behavior and cognition. This series covers both practical and theoretical aspects of human factors in transportation, with an emphasis on their interaction.

The series is intended as a forum for researchers and engineers interested in how people function within transportation systems. All modes of transportation are relevant, and all human factors and ergonomic efforts that have explicit implications for transportation systems fall within a series purview. Analytic efforts are important to link theory and data. The level of analysis can be as small as one person, or international in scope. Empirical data can be from a broad range of methodologies, including laboratory research, simulator studies, test tracks, operational tests, fieldwork, design reviews, or surveys. This broad scope is intended to maximize the utility of the series for readers with diverse backgrounds.

I expect the series to be useful for professionals in the disciplines of human factors, ergonomics, transportation engineering, experimental psychology, cognitive science, sociology, and safety engineering. It is intended to appeal to the transportation specialist in industry, government, or academics, as well as the researcher in need of a testbed for new ideas about the interface between people and complex systems.

This book, while focusing on human error, offers a systems approach that is particularly welcome in transportation human factors. A major goal of this book series is to link theory and practice of human factors. The author is to be commended for asking questions that not only link theory and practice, but force the reader to evaluate classes of theory as applied to human factors. Traditional information theory approaches, derived from the limited-channel model that has formed the original basis for theoretical work in human factors, are held up to scrutiny. Newer approaches such as situational awareness, that spring from deficiencies in the information theory model, are criticized as being only folk models that lack scientific rigor. I hope this book engenders a vigorous debate as to what kinds of theory best serve the science of human factors. Although the ten questions offered here form a basis for debate, there are more than ten possible answers. Forthcoming books in this series will continue to search for these answers by blending practical and theoretical perspectives in transportation human factors.

Author Note

Sidney Dekker is Professor of Human Factors at Lund University, Sweden. He received an M.A. in organizational psychology from the University of Nijmegen and an M.A. in experimental psychology from Leiden University, both in the Netherlands. He gained his Ph.D. in Cognitive Systems Engineering from The Ohio State University.

He has previously worked for the Public Transport Cooperation in Melbourne, Australia; the Massey University School of Aviation, New Zealand; and British Aerospace. His specialties and research interests are human error, accident investigations, field studies, representation design, and automation. He has some experience as a pilot, type trained on the DC-9 and Airbus A340. His previous books include *The Field Guide to Human Error Investigations* (2002).

Chapter 1

Was It Mechanical Failure
or Human Error?

These are exciting and challenging times for human factors and system safety. And there are indications that we may not be entirely well equipped for them. There is an increasing recognition that mishaps (a commercial aircraft accident, a Space Shuttle disaster) are inextricably linked to the functioning of surrounding organizations and institutions. The operation of commercial airliners or Space Shuttles or passenger ferries spawns vast networks of organizations to support it, to advance and improve it, to control and regulate it. Complex technologies cannot exist without these organizations and institutions—carriers, regulators, government agencies, manufacturers, subcontractors, maintenance facilities, training outfits—that, in principle, are designed to protect and secure their operation. Their very mandate boils down to not having accidents happen. Since the 1978 nuclear accident at Three Mile Island, however, people increasingly realize that the very organizations meant to keep a technology safe and stable (human operators, regulators, management, maintenance) are actually among the major contributors to breakdown. Sociotechnical failures are impossible without such contributions.

Despite this growing recognition, human factors and system safety relies on a vocabulary based on a particular conception of the natural sciences, derived from its roots in engineering and experimental psychology. This vocabulary, the subtle use of metaphors, images, and ideas is more and more at odds with the interpretative demands posed by modern organizational accidents. The vocabulary expresses a worldview (perhaps) appropriate for technical failures, but incapable of embracing and penetrating the relevant areas of sociotechnical failures—those failures that involve the intertwined effects of technology and the organized social complexity surrounding its use. Which is to say, most failures today.

Any language, and the worldview it mediates, imposes limitations on our understanding of failure. Yet these limitations are now becoming increasingly evident and pressing. With growth in system size and complexity, the nature of accidents is changing (system accidents, sociotechnical failures). Resource scarcity and competition mean that systems incrementally push their operations toward the edges of their safety envelopes. They have to do this in order to remain successful in their dynamic environments. Commercial returns at the boundaries are greater, but the difference between having and not having an accident are up to stochastics more than available margins. Open systems are continually adrift within their safety envelopes, and the processes that drive such migration are not easy to recognize or control, nor is the exact location of the boundaries. Large, complex systems seem capable of acquiring a hysteresis, an obscure will of their own, whether they are drifting towards greater resilience or towards the edges of failure. At the same time, the fast pace of technological change creates new types of hazards, especially those that come with increased reliance on computer technology. Both engineered and social systems (and their interplay) rely to an ever greater extent on information technology. Although computational speed and access to information would seem a safety advantage in principle, our ability to make sense of data is not at all keeping pace with our ability to collect and generate it. By knowing more, we may actually know a lot less. Managing safety by numbers (incidents, error counts, safety threats), as if safety is just another index of a Harvard business model, can create a false impression of rationality and managerial control. It may ignore higher order variables that could unveil the true nature and direction of system drift. It may also come at the cost of deeper understandings of real sociotechnical functioning.

DECONSTRUCTION, DUALISM, AND STRUCTURALISM

What is that language, then, and the increasingly obsolete technical worldview it represents? Its defining characteristics are deconstruction, dualism, and structuralism. *Deconstruction* means that a system's functioning can be understood exhaustively by studying the arrangement and interaction of its constituent parts. Scientists and engineers typically look at the world this way. Accident investigations deconstruct too. In order to rule out mechanical failure, or to locate the offending parts, accident investigators speak of "reverse engineering." They recover parts from the rubble and reconstruct them into a whole again, often quite literally. Think of the TWA800 Boeing

747 that exploded in midair after takeoff from New York's Kennedy airport in 1998. It was recovered from the Atlantic Ocean floor and painstakingly pieced back together—if heavily scaffolded—in a hangar. With the puzzle as complete as possible, the broken part(s) should eventually get exposed, allowing investigators to pinpoint the source of the explosion. Accidents are puzzling wholes. But it continues to defy sense, it continues to be puzzling only when the functioning (or non functioning) of its parts fail to explain the whole. The part that caused the explosion, that ignited it, was never actually pinpointed. This is what makes the TWA800 investigation scary. Despite one of the most expensive reconstructions in history, the reconstructed parts refused to account for the behavior of the whole. In such a case, a frightening, uncertain realization creeps into the investigator corps and into industry. A whole failed without a failed part. An accident happened without a cause; no cause—nothing to fix, nothing to fix—it could happen again tomorrow, or today.

The second defining characteristic is dualism. *Dualism* means that there is a distinct separation between material and human cause—between human error or mechanical failure. In order to be a good dualist, you of course have to deconstruct: You have to disconnect human contributions from mechanical contributions. The rules of the International Civil Aviation Organization that govern aircraft accident investigators prescribe exactly that. They force accident investigators to separate human contributions from mechanical ones. Specific paragraphs in accident reports are reserved for tracing the potentially broken human components. Investigators explore the anteceding 24- and 72-hour histories of the humans who would later be involved in a mishap. Was there alcohol? Was there stress? Was there fatigue? Was there a lack of proficiency or experience? Were there previous problems in the training or operational record of these people? How many flight hours did the pilot really have? Were there other distractions or problems? This investigative requirement reflects a primeval interpretation of human factors, an aeromedical tradition where human error is reduced to the notion of "fitness for duty." This notion has long been overtaken by developments in human factors towards the study of normal people doing normal work in normal workplaces (rather than physiologically or mentally deficient miscreants), but the overextended aeromedical model is retained as a kind of comforting positivist, dualist, deconstructive practice. In the fitness-for-duty paradigm, sources of human error must be sought in the hours, days or years before the accident, when the human component was already bent and weakened and ready to break. Find the part of the human that was missing or deficient, the "unfit part," and the human part will carry the interpretative load of the accident. Dig into recent history, find the deficient pieces and put the puzzle together: deconstruction, reconstruction, and dualism.

The third defining characteristic of the technical worldview that still governs our understanding of success and failure in complex systems is structuralism. The language we use to describe the inner workings of successful and failed systems is a language of structures. We speak of layers of defense, of holes in those layers. We identify the "blunt ends" and "sharp ends" of organizations and try to capture how one has effects on the other. Even safety culture gets treated as a structure consisting of other building blocks. How much of a safety culture an organization has depends on the routines and components it has in place for incident reporting (this is measurable), to what extent it is just in treating erring operators (this is more difficult to measure, but still possible), and what linkages it has between its safety functions and other institutional structures. A deeply complex social reality is thus reduced to a limited number of measurable components. For example, does the safety department have a direct route to highest management? What is the reporting rate compared to other companies?

Our language of failures is also a language of mechanics. We describe accident trajectories, we seek causes and effects, and interactions. We look for initiating failures, or triggering events, and trace the successive domino-like collapse of the system that follows it. This worldview sees sociotechnical systems as machines with parts in a particular arrangement (blunt vs. sharp ends, defenses layered throughout), with particular interactions (trajectories, domino effects, triggers, initiators), and a mix of independent or intervening variables (blame culture vs. safety culture). This is the worldview inherited from Descartes and Newton, the worldview that has successfully driven technological development since the scientific revolution half a millennium ago. The worldview, and the language it produces, is based on particular notions of natural science, and exercises a subtle but very powerful influence on our understanding of sociotechnical success and failure today. As it does with most of Western science and thinking, it pervades and directs the orientation of human factors and system safety.

Yet language, if used unreflectively, easily becomes imprisoning. Language expresses but also determines what we can see and how we see it. Language constrains how we construct reality. If our metaphors encourage us to model accident chains, then we will start our investigation by looking for events that fit in that chain. But which events should go in? Where should we start? As Nancy Leveson (2002) pointed out, the choice of which events to put in is arbitrary, as are the length, the starting point and level of detail of the chain of events. What, she asked, justifies assuming that initiating events are mutually exclusive, except that it simplifies the mathematics of the failure model? These aspects of technology, and of operating it, raise questions about the appropriateness of the dualist, deconstructed, structuralist model that dominates human factors and system safety. In its place we may seek a true systems view, which not only maps the structural deficiencies behind individual human errors (if indeed it does that at all), but that appreciates the organic, ecological adaptability of complex sociotechnical systems.

Looking for Failures to Explain Failures

Our most entrenched beliefs and assumptions often lie locked up in the simplest of questions. The question about mechanical failure or human error is one of them. Was the accident caused by mechanical failure or by human error? It is a stock question in the immediate aftermath of a mishap. Indeed, it seems such a simple, innocent question. To many it is a normal question to ask: If you have had an accident, it makes sense to find out what broke. The question, however, embodies a particular understanding of how accidents occur, and it risks confining our causal analysis to that understanding. It lodges us into a fixed interpretative repertoire. Escaping from this repertoire may be difficult. It sets out the questions we ask, provides the leads we pursue and the clues we examine, and determines the conclusions we will eventually draw. Which components were broken? Was it something engineered, or some human? How long had the component been bent or otherwise deficient? Why did it eventually break? What were the latent factors that conspired against it? Which defenses had eroded?

These are the types of questions that dominate inquiries in human factors and system safety today. We organize accident reports and our discourse about mishaps around the struggle for answers to them. Investigations turn up broken mechanical components (a failed jackscrew in the vertical trim of an Alaska Airlines MD-80, perforated heat tiles in the Columbia Space Shuttle), underperforming human components (e.g., breakdowns in crew resource management, a pilot who has a checkered training record), and cracks in the organizations responsible for running the system (e.g., weak organizational decision chains, deficient maintenance, failures in regulatory oversight). Looking for failures—human, mechanical, or organizational—in order to explain failures is so common-sensical that most investigations never stop to think whether these are indeed the right clues to pursue. That failure is caused by failure is prerational—we do not consciously consider it any longer as a question in the decisions we make about where to look and what to conclude.

Here is an example. A twin-engined Douglas DC-9-82 landed at a regional airport in the Southern Highlands of Sweden in the summer of 1999. Rainshowers had passed through the area earlier, and the runway was still wet. While on approach to the runway, the aircraft got a slight tailwind, and after touchdown the crew had trouble slowing down. Despite increasing crew efforts to brake, the jet overran the runway and ended up in a field a few hundred feet from the threshold. The 119 passengers and crew onboard were unhurt. After coming to a standstill, one of the pilots made his way out of the aircraft to check the brakes. They were stone cold. No wheel braking had occurred at all. How could this have happened? Investigators found no mechanical failures on the aircraft. The braking systems were fine.

Instead, as the sequence of events was rolled back in time, investigators realized that the crew had not armed the aircraft's ground spoilers before landing. Ground spoilers help a jet aircraft brake during rollout, but they need to be armed before they can do their work. Arming them is the job of the pilots, and it is a before-landing checklist item and part of the procedures that both crewmembers are involved in. In this case, the pilots forgot to arm the spoilers. "Pilot error," the investigation concluded.

Or actually, they called it "Breakdowns in CRM (Crew Resource Management)" (Statens Haverikommision, 2000, p. 12), a more modern, more euphemistic way of saying "pilot error." The pilots did not coordinate what they should have; for some reason they failed to communicate the required configuration of their aircraft. Also, after landing, one of the crew members had not called "Spoilers!" as the procedures dictated. This could, or should, have alerted the crew to the situation, but it did not happen. Human errors had been found. The investigation was concluded.

"Human error" is our default when we find no mechanical failures. It is a forced, inevitable choice that fits nicely into an equation, where human error is the inverse of the amount of mechanical failure. Equation 1 shows how we determine the ratio of causal responsibility:

$$\text{human error} = f(1 - \text{mechanical failure}) \qquad (1)$$

If there is no mechanical failure, then we know what to start looking for instead. In this case, there was no mechanical failure. Equation 1 came out as a function of 1 minus 0. The human contribution was 1. It was human error, a breakdown of CRM. Investigators found that the two pilots onboard the MD-80 were actually both captains, and not a captain and a copilot as is usual. It was a simple and not altogether uncommon scheduling fluke, a stochastic fit flying onboard the aircraft since that morning. With two captains on a ship, responsibilities risk getting divided unstably and incoherently. Division of responsibility easily leads to its abdication. If it is the role of the copilot to check that the spoilers are armed, and there is no copilot, the risk is obvious. The crew was in some sense "unfit," or at least prone to breaking down. It did (there was a "breakdown of CRM"). But what does this explain? These are processes that themselves require an explanation, and they may be leads that go cold anyway. Perhaps there is a much deeper reality lurking beneath the prima facie particulars of such an incident, a reality where machinistic and human cause are much more deeply intertwined than our formulaic approaches to investigations allow us to grasp. In order to get better glimpses of this reality, we first have to turn to dualism. It is dualism that lies at the heart of the choice between human error and mechanical failure. We take a brief peek at its past and confront it with the unstable, uncertain empirical encounter of an unarmed spoilers case.

Dualism Destitute

The urge to separate human cause from machinic cause is something that must have puzzled even the early nascent human factors tinkerers. Think of the fiddling with World War II cockpits that had identical control switches for a variety of functions. Would a flap-like wedge on the flap handle and a wheel-shaped cog on the gear lever avoid the typical confusion between the two? Both common sense and experience said "yes." By changing something in the world, human factors engineers (to the extent that they existed already) changed something in the human. By toying with the hardware that people worked with, they shifted the potential for correct versus incorrect action, but only the potential. For even with functionally shaped control levers, some pilots, in some cases, still got them mixed up. At the same time, pilots did not always get identical switches mixed up. Similarly, not all crews consisting of two captains fail to arm the spoilers before landing. Human error, in other words, is suspended, unstably, somewhere between the human and the engineered interfaces. The error is neither fully human, nor fully engineered. At the same time, mechanical "failures" (providing identical switches located next to one another) get to express themselves in human action. So if a confusion between flaps and gear occurs, then what is the cause? Human error or mechanical failure? You need both to succeed; you need both to fail. Where one ends and the other begins is no longer so clear. One insight of early human factors work was that machinistic feature and human action are intertwined in ways that resist the neat, dualist, deconstructed disentanglement still favored by investigations (and their consumers) today.

DUALISM AND THE SCIENTIFIC REVOLUTION

The choice between human cause and material cause is not just a product of recent human factors engineering or accident investigations. The choice is firmly rooted in the Cartesian–Newtonian worldview that governs much of our thinking to this day, particularly in technologically dominated professions such as human factors engineering and accident investigation. Isaac Newton and René Descartes were two of the towering figures in the Scientific Revolution between 1500 and 1700 A.D. which produced a dramatic shift in worldview, as well as profound changes in knowledge and in ideas on how to acquire and test knowledge. Descartes proposed a sharp distinction between what he called *res cogitans*, the realm of mind and thought, and *res extensa*, the realm of matter. Although Descartes admitted to some interaction between the two, he insisted that mental and physical phenomena cannot be understood by reference to each other. Problems that occur in either realm require entirely separate approaches and different concepts to solve them. The notion of separate mental and material worlds became known as dualism and its implications can be recognized in much of what we think and do today. According to Descartes, the mind is outside of the physical order of matter and is in no way derived from it. The choice between human error and mechanical failure is such a dualist choice: According to Cartesian logic, human error cannot be derived from material things. As we will see, this logic does not hold up well—in fact, on closer inspection, the entire field of human factors is based on its abrogation.

Separating the body from the soul, and subordinating the body to the soul, not only kept Descartes out of trouble with the Church. His dualism, his division between mind and matter, addressed an important philosophical problem that had the potential of holding up scientific, technological, and societal progress: What is the link between mind and matter, between the soul and the material world? How could we, as humans, take control of and remake our physical world as long as it was indivisibly allied to or even synonymous with an irreducible, eternal soul? A major aim during the 16th- and 17th-century Scientific Revolution was to see and understand (and become able to manipulate) the material world as a controllable, predictable, programmable machine. This required it to be seen as nothing but a machine: No life, no spirit, no soul, no eternity, no immaterialism, no unpredictability. Descartes' *res extensa*, or material world, answered to just that concern. The *res extensa* was described as working like a machine, following mechanical rules and allowing explanations in terms of the arrangement and movement of its constituent parts. Scientific progress became easier because of what it excluded. What the Scientific Revolution required, Descartes' disjunction provided. Nature became a perfect machine, governed by mathematical laws that were increasingly within the grasp of human understanding and control, and away from things humans cannot control. Newton, of course, is the father of many of the laws that still govern our understanding of the universe today. His third law of motion, for example, lies at the basis of our presumptions about cause and effect, and causes of accidents: For every action there is an equal and opposite reaction. In other words, for each cause there is an equal effect, or rather, for each effect there must be an equal cause. Such a law, though applicable to the release and transfer of energy in mechanical systems, is misguiding and disorienting when applied to sociotechnical failures, where the small banalities and subtleties of normal work done by normal people in normal organizations can slowly degenerate into enormous disasters, into disproportionately huge releases of energy. The cause–consequence equivalence dictated by Newton's third law of motion is quite inappropriate as a model for organizational accidents.

Attaining control over a material world was critically important for people five hundred years ago. The inspiration and fertile ground for the ideas of Descartes and Newton can be understood against the background of their time. Europe was emerging from the Middle Ages—fearful and fateful times, where life spans were cut short by wars, disease, and epidemics. We should not underestimate anxiety and apprehension about humanity's ability to make it at all against these apocryphal odds. After the Plague, the population of Newton's native England, for example, took until 1650 to recover to the level of 1300. People were at the mercy of ill-understood and barely controllable forces. In the preceding millennium, piety, prayer, and penitence were among the chief mechanisms through which people could hope to attain some kind of sway over ailment and disaster.

The growth of insight produced by the Scientific Revolution slowly began to provide an alternative, with measurable empirical success. The Scientific Revolution provided new means for controlling the natural world. Telescopes and microscopes gave people new ways of studying components that had thus far been too small or too far away for the naked eye to see, cracking open a whole new view on the universe and for the first time revealing causes of phenomena hitherto ill understood. Nature was not a monolithic, inescapable bully, and people were no longer just on the receiving, victimized end of its vagaries. By studying it in new ways, with new instruments, nature could be decomposed, broken into smaller bits, measured, and, through all of that, better understood and eventually controlled. Advances in mathematics (geometry, algebra, calculus) generated models that could account for and begin to predict newly discovered phenomena in, for example, medicine and astronomy. By discovering some of the building blocks of life and the universe, and by developing mathematical imitates of their functioning, the Scientific Revolution reintroduced a sense of predictability and control that had long been dormant during the Middle Ages. Humans could achieve dominance and preeminence over the vicissitudes and unpredictabilities of nature. The route to such progress would come from measuring, breaking down (known variously today as reducing, decomposing, or deconstructing) and mathematically modeling the world around us—to subsequently rebuild it on our terms.

Measurability and control are themes that animated the Scientific Revolution, and they resonate strongly today. Even the notions of dualism (material and mental worlds are separate) and deconstruction (larger wholes can be explained by the arrangement and interaction of their constituent lower level parts) have long outlived their initiators. The influence of Descartes is judged so great in part because he wrote in his native tongue, rather than in Latin, thereby presumably widening access and popular exposure to his thoughts. The mechanization of nature spurred by his dualism, and Newton's and others' enormous mathematical advances, heralded centuries of unprecedented scientific progress, economic growth, and engineering success. As Fritjof Capra (1982) put it, NASA would not have been able to put a man on the moon without René Descartes.

The heritage, however, is definitely a mixed blessing. Human factors and systems safety is stuck with a language, with metaphors and images that emphasize structure, components, mechanics, parts and interactions, cause and effect. While giving us initial direction for building safe systems and for finding out what went wrong when it turns out we have not, there are limits to the usefulness of this inherited vocabulary. Let us go back to that summer day of 1999 and the MD-80 runway overrun. In good Cartesian–Newtonian tradition, we can begin by opening up the aircraft a bit more, picking apart the various components and procedures to see how they interact, second by second. Initially we will be met with resounding empirical success—as indeed Descartes and Newton frequently were. But when we want to recreate the whole on the basis of the parts we find, a more troubling reality swims into view: It does not go well together anymore. The neat, mathematically pleasing separation between human and mechanical cause, between social and structural issues, has blurred. The whole no longer seems a linear function of the sum of the parts. As Scott Snook (2000) explained it, the two classical Western scientific steps of analytic reduction (the whole into parts) and inductive synthesis (the parts back into a whole again) may seem to work, but simply putting the parts we found back together does not capture the rich complexity hiding inside and around the incident. What is needed is a holistic, organic integration. What is perhaps needed is a new form of analysis and synthesis, sensitive to the total situation of organized sociotechnical activity. But first let us examine the analytical, componential story.

SPOILERS, PROCEDURES, AND HYDRAULIC SYSTEMS

Spoilers are those flaps that come up into the airstream on the topside of the wings after an aircraft has touched down. Not only do they help brake the aircraft by obstructing the airflow, they also cause the wing to lose the ability to create lift, forcing the aircraft's weight onto the wheels. Extension of the ground spoilers also triggers the automatic braking system on the wheels: The more weight the wheels carry, the more effective their braking becomes. Before landing, pilots select the setting they wish on the automatic wheel-braking system (minimum, medium or maximum), depending on runway length and conditions. After landing, the automatic wheel-braking system will slow down the aircraft without the pilot having to do anything, and without letting the wheels skid or lose traction. As a third mechanism for slowing down, most jet aircraft have thrust reversers, which redirect the outflow of the jet engines into the oncoming air, instead of out toward the back.

In this case, no spoilers came out, and no automatic wheel braking was triggered as a result. While rolling down the runway, the pilots checked the setting of the automatic braking system multiple times to ensure it was armed and even changed its setting to maximum as they saw the end of the runway coming up. But it would never engage. The only remaining mechanism for slowing down the aircraft were the thrust reversers. Thrust reversers, however, are most effective at high speeds. By the time the pilots noticed that they were not going to make it before the end of the runway, the speed was already quite low (they ended up going into the field at 10–20 knots) and thrust reversers no longer had much immediate effect. As the jet was going over the edge of the runway, the captain closed the reversers and steered somewhat to the right in order to avoid obstacles.

How are spoilers armed? On the center pedestal, between the two pilots, are a number of levers. Some are for the engines and thrust reversers, one is for the flaps, and one is for the spoilers. In order to arm the ground spoilers, one of the pilots needs to pull the lever upward. The lever goes up by about one inch and sits there, armed until touchdown. When the system senses that the aircraft is on the ground (which it does in part through switches in the landing gear), the lever will come back automatically and the spoilers come out. Asaf Degani, who studied such procedural problems extensively, has called the spoiler issue not one of human error, but one of timing (e.g., Degani, Heymann, & Shafto, 1999). On this aircraft, as on many others, the spoilers should not be armed before the landing gear has been selected down and is entirely in place. This has to do with the switches that can tell when the aircraft is on the ground. These are switches that compress as the aircraft's weight settles onto the wheels, but not only then. There is a risk in this type of aircraft that the switch in the nose gear will even compress as the landing gear is coming out of its bays. This can happen because the nose gear folds out into the oncoming airstream. As the nose gear is coming out and the aircraft is slicing through the air at 180 knots, the sheer wind force can compress the nose gear, activate the switch, and subsequently risk extending the ground spoilers (if they had been armed). This is not a good idea: The aircraft would have trouble flying with ground spoilers out. Hence the requirement: The landing gear needs to be all the way out, pointing down. Only when there is no more risk of aerodynamic switch compression can the spoilers be armed. This is the order of the before-landing procedures:

Gear down and locked.

Spoilers armed.

Flaps FULL.

On a typical approach, pilots select the landing gear handle down when the so-called glide slope comes alive: when the aircraft has come within range of the electronic signal that will guide it down to the runway. Once the landing gear is out, spoilers must be armed. Then, once the aircraft captures that glide slope (i.e., it is exactly on the electronic beam) and starts to descend down the approach to the runway, flaps need to be set to FULL (typically 40°). Flaps are other devices that extend from the wing, changing the wing size and shape. They allow aircraft to fly more slowly for a landing. This makes the procedures conditional on context. It now looks like this:

Gear down and locked (when glide slope live).

Spoilers armed (when gear down and locked).

Flaps FULL (when glide slope captured).

But how long does it take to go from "glide slope live" to "glide slope capture"? On a typical approach (given the airspeed) this takes about 15 seconds. On a simulator, where training takes place, this does not create a problem. The whole gear cycle (from gear lever down to the "gear down and locked" indication in the cockpit) takes about 10 seconds. That leaves 5 seconds for arming the spoilers, before the crew needs to select flaps FULL (the next item in the procedures). In the simulator, then, things look like this:

At t = 0   Gear down and locked (when glide slope live).

At t + 10   Spoilers armed (when gear down and locked).

At t + 15   Flaps FULL (when glide slope captured).

But in real aircraft, the hydraulic system (which, among other things, extends the landing gear) is not as effective as it is on a simulator. The simulator, of course, only has simulated hydraulic aircraft systems, modeled on how the aircraft is when it has flown zero hours, when it is sparkling new, straight out of the factory. On older aircraft, it can take up to half a minute for the gear to cycle out and lock into place. This makes the procedures look like this:

At t = 0   Gear down and locked (when glide slope live).

At t + 30   Spoilers armed (when gear down and locked).

BUT! at t + 15   Flaps FULL (when glide slope captured).

In effect, then, the "flaps" item in the procedures intrudes before the "spoilers" item. Once the "Flaps" item is completed and the aircraft is descending towards the runway, it is easy to go down the procedures from there, taking the following items. Spoilers never get armed. Their arming has tumbled through the cracks of a time warp. An exclusive claim to human error (or CRM breakdown) becomes more difficult to sustain against this background. How much human error was there, actually? Let us remain dualist for now and revisit Equation 1. Now apply a more liberal definition of *mechanical failure*. The nose gear of the actual aircraft, fitted with a compression switch, is designed so that it folds out into the wind while still airborne. This introduces a systematic mechanical vulnerability that is tolerated solely through procedural timing (a known leaky mechanism against failure): first the gear, then the spoilers. In other words, "gear down and locked" is a mechanical prerequisite for spoiler arming, but the whole gear cycle can take longer than there is room in the procedures and the timing of events driving their application. The hydraulic system of the old jet does not pressurize as well: It can take up to 30 seconds for a landing gear to cycle out. The aircraft simulator, in contrast, does the same job inside of 10 seconds, leaving a subtle but substantive mechanical mismatch. One work sequence is introduced and rehearsed in training, whereas a delicately different one is necessary for actual operations. Moreover, this aircraft has a system that warns if the spoilers are not armed on takeoff, but it does not have a system for warning that the spoilers are not armed on approach. Then there is the mechanical arrangement in the cockpit. The armed spoiler handle looks different from the unarmed one by only one inch and a small red square at the bottom. From the position of the right-seat pilot (who needs to confirm their arming), this red patch is obscured behind the power levers as these sit in the typical approach position. With so much mechanical contribution going around (landing gear design, eroded hydraulic system, difference between simulator and real aircraft, cockpit lever arrangement, lack of spoiler warning system on approach, procedure timing) and a helping of scheduling stochastics (two captains on this flight), a whole lot more mechanical failure could be plugged into the equation to rebalance the human contribution.

But that is still dualist. When reassembling the parts that we found among procedures, timing, mechanical erosion, design trade-offs, we can begin to wonder where mechanical contributions actually end, and where human contributions begin. The border is no longer so clear. The load imposed by a wind of 180 knots on the nose wheel is transferred onto a flimsy procedure: first the gear, then the spoilers. The nose wheel, folding out into the wind and equipped with a compression switch, is incapable of carrying that load and guaranteeing that spoilers will not extend, so a procedure gets to carry the load instead. The spoiler lever is placed in a way that makes verification difficult, and a warning system for unarmed spoilers is not installed. Again, the error is suspended, uneasily and unstably, between human intention and engineered hardware—it belongs to both and to neither uniquely. And then there is this: The gradual wear of a hydraulic system is not something that was taken into account during the certification of the jet. An MD-80 with an anemic hydraulic system that takes more than half a minute to get the whole gear out, down, and locked, violating the original design requirement by a factor of three, is still considered airworthy. The worn hydraulic system cannot be considered a mechanical failure. It does not ground the jet. Neither do the hard-to-verify spoiler handle or lack of warning system on approach. The jet was once certified as airworthy with or without all of that. That there is no mechanical failure, in other words, is not because there are no mechanical issues. There is no mechanical failure because social systems, made up of manufacturers, regulators, and prospective operators—undoubtedly shaped by practical concerns and expressed through situated engineering judgment with uncertainty about future wear—decided that there would not be any (at least not related to the issues now identified in an MD-80 overrun). Where does mechanical failure end and human error begin? Dig just deeply enough and the question becomes impossible to answer.

RES EXTENSA AND RES COGITANS, OLD AND NEW

Separating *res extensa* away from *res cogitans*, like Descartes did, is artificial. It is not the result of natural processes or conditions, but rather an imposition of a worldview. This worldview, though initially accelerating scientific progress, is now beginning to seriously hamper our understanding. In modern accidents, machinistic and human causes are blurred. The disjunction between material and mental worlds, and the requirement to describe them differently and separately, are debilitating our efforts to understand sociotechnical success and failure.

The distinction between the old and new views of human error, which was also made the earlier *Field Guide to Human Error Investigations* (Dekker, 2002) actually rides roughshod over these subtleties. Recall how the investigation into the runway overrun incident found "breakdowns in CRM" as a causal factor. This is old-view thinking. Somebody, in this case a pilot, or rather a crew of two pilots, forgot to arm the spoilers. This was a human error, an omission. If they had not forgotten to arm the spoilers, the accident would not have happened, end of story. But such an analysis of failure does not probe below the immediately visible surface variables of a sequence of events. As Perrow (1984) puts it, it judges only where people should have zigged instead of zagged. The old view of human error is surprisingly common. In the old view, error—by any other name (e.g., complacency, omission, breakdown of CRM)—is accepted as a satisfactory explanation. This is what the new view of human error tries to avoid. It sees human error as a consequence, as a result of failings and problems deeper inside the systems in which people work. It resists seeing human error as the cause. Rather than judging people for not doing what they should have done, the new view presents tools for explaining why people did what they did. Human error becomes a starting point, not a conclusion. In the spoiler case, the error is a result of design trade-offs, mechanical erosion, procedural vulnerabilities, and operational stochastics. Granted, the commitment of the new view is to resist neat, condensed versions in which a human choice or a failed mechanical part led the whole structure onto the road to perdition. The distinction between the old and new views is important and needed. Yet even in the new view the error is still an effect, and effects are the language of Newton. The new view implicitly acknowledges the existence, the reality of error. It sees error as something that is out there, in the world, and caused by something else, also out there in the world. As the next chapters show, such a (naively) realist position is perhaps untenable.

Recall how the Newtonian–Cartesian universe consists of wholes that can be explained and controlled by breaking them down into constituent parts and their interconnections (e.g., humans and machines, blunt ends and sharp ends, safety cultures and blame cultures). Systems are made up of components, and of mechanical-like linkages between those components. This lies at the source of the choice between human and material cause (is it human error or mechanical failure?). It is Newtonian in that it seeks a cause for any observed effect, and Cartesian in its dualism. In fact, it expresses both Descartes' dualism (either mental or material: You cannot blend the two) and the notion of decomposition, where lower order properties and interactions completely determine all phenomena. They are enough; you need no more. Analyzing which building blocks go into the problem, and how they add up, is necessary and sufficient for understanding why the problem occurs. Equation 1 is a reflection of the assumed explanatory sufficiency of lower order properties. Throw in the individual contributions, and the answer to why the problem occurred rolls out. An aircraft runway overrun today can be understood by breaking the contributions down into human and machine causes, analyzing the properties and interactions of each and then reassembling it back into a whole. "Human error" turns up as the answer. If there are no material contributions, the human contribution is expected to carry the full explanatory load.

As long as progress is made using this worldview, there is no reason to question it. In various corners of science, including human factors, many people still see no reason to do so. Indeed, there is no reason that structuralist models cannot be imposed on the messy interior of sociotechnical systems. That these systems, however, reveal machine-like properties (components and interconnections, layers and holes) when we open them up post mortem does not mean that they are machines, or that they, in life, grew and behaved like machines. As Leveson (2002) pointed out, analytic reduction assumes that the separation of a whole into constituent parts is feasible, that the subsystems operate independently, and that analysis results are not distorted by taking the whole apart. This in turn implies that the components are not subject to feedback loops and other nonlinear interactions and that they are essentially the same when examined singly as when they are playing their part in the whole. Moreover, it assumes that the principles governing the assembly of the components into the whole are straightforward; the interactions among components are simple enough that they can be considered separate from the behavior of the whole.

Are these assumptions valid when we try to understand system accidents? The next chapters give us cause to think. Take for example the conundrum of drift into failure and the elusive nature of accidents that happen beyond a 10^{-7} safety level. Those accidents do not happen just because of component failures, yet our mechanistic models of organizational or human functioning can never capture the organic, relational processes that gradually nudge a sociotechnical system to the edge of breakdown. Looking at component failures, such as the "human errors" that many popular error categorization methods look for, may be fraudulent in its illusion of what it tells us about safety and risk in complex systems. There is a growing consensus that our current efforts and models will be incapable of breaking the asymptote, the level-off, in our progress on safety beyond 10^{-7}. Is the structuralist, mechanistic view of sociotechnical systems, where we see components and linkages and their failures, still appropriate for making real progress?

Chapter 2

Why Do Safe Systems Fail?

Accidents actually do not happen very often. Most transportation systems in the developed world are safe or even ultra-safe. Their likelihood of a fatal accident is less than 10^{-7}, which means a one-out-of-10,000,000 chance of death, serious loss of property or environmental or economic devastation (Amalberti, 2001). At the same time, this appears to be a magical frontier: No transportation system has figured out a way of becoming even safer. Progress on safety beyond 10^{-7} is elusive. As René Amalberti has pointed out, linear extensions of current safety efforts (incident reporting, safety and quality management, proficiency checking, standardization and proceduralization, more rules and regulations) seem of little use in breaking the asymptote, even if they are necessary to sustain the 10^{-7} safety level.

More intriguingly still, the accidents that happen at this frontier appear to be of a type that is difficult to predict using the logic that governs safety thinking up to 10^{-7}. It is here that the limitations of a structuralist vocabulary become most apparent. Accident models that rely largely on failures, holes, violations, deficiencies, and flaws can have a difficult time accommodating accidents that seem to emerge from (what looks to everybody like) normal people doing normal work in normal organizations. Yet the mystery is that in the hours, days, or even years leading up to an accident beyond 10^{-7}, there may be few reportworthy failures or noteworthy organizational deficiencies. Regulators as well as insiders typically do not see people violating rules, nor do they discover other flaws that would give cause to shut down or seriously reconsider operations. If only it were that easy. And up to 10^{-7} it probably is. But when failures, serious failures, are no longer preceded by serious failures, predicting accidents becomes a lot more difficult.

17

And modeling them with the help of mechanistic, structuralist notions may be of little help.

The greatest residual risk in today's safe sociotechnical systems is the drift into failure. Drift into failure is about a slow, incremental movement of systems operations toward the edge of their safety envelope. Pressures of scarcity and competition typically fuel this drift. Uncertain technology and incomplete knowledge about where the boundaries actually are, result in people not stopping the drift or even seeing it. The 2000 Alaska Airlines 261 accident is highly instructive in this sense. The MD-80 crashed into the ocean off California after the trim system in its tail snapped. On the surface, the accident seems to fit a simple category that has come to dominate recent accident statistics: mechanical failures as a result of poor maintenance: A single component failed because people did not maintain it well. Indeed, there was a catastrophic failure of a single component. A mechanical failure, in other words. The break instantly rendered the aircraft uncontrollable and sent it plummeting into the Pacific. But such accidents do not happen just because somebody suddenly errs or something suddenly breaks: There is supposed to be too much built-in protection against the effects of single failures. What if these protective structures themselves contribute to drift, in ways inadvertent, unforeseen, and hard to detect? What if the organized social complexity surrounding the technological operation, all the maintenance committees, working groups, regulatory interventions, approvals, and manufacturer inputs, that all intended to protect the system from breakdown, actually helped to set its course to the edge of the envelope?

Since Barry Turner's 1978 *Man-Made Disasters*, we know explicitly that accidents in complex, well-protected systems are incubated. The potential for an accident accumulates over time, but this accumulation, this steady slide into disaster, generally goes unrecognized by those on the inside and even those on the outside. So Alaska 261 is not just about a mechanical failure, even if that is what many people would like to see as the eventual outcome (and proximal cause of the accident). Alaska 261 is about uncertain technology, about gradual adaptations, about drift into failure. It is about the inseparable, mutual influences of mechanical and social worlds, and it puts the inadequacy of our current models in human factors and system safety on full display.

JACKSCREWS AND MAINTENANCE JUDGMENTS

In Alaska 261, the drift toward the accident that happened in 2000 had begun decades earlier. It reaches back into the very first flights of the 1965 Douglas DC-9 that preceded the MD-80 type. Like (almost) all aircraft, this

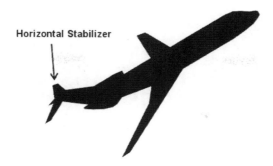

FIG. 2.1.   The location of the horizontal stabilizer on an MD-80 type aircraft.

type has a horizontal stabilizer (or tailplane, a small wing) at the back that helps direct the lift created by wings. It is this little tailplane that keeps an aircraft's nose up: Without it, controlled flight is not possible (see Fig. 2.1). The tailplane itself can angle up or down in order to pitch the nose up or down (and consequently make the aircraft go up or down). In most aircraft, the trim system can be driven by the autopilot and by crew inputs. The tailplane is hinged at the back, whereas the front end arcs up or down (it also has control surfaces at the back that are connected to the crew's control column in the cockpit, but those are not the issue here). Pushing the front end of the horizontal stabilizer up or down is done through a rotating jackscrew and a nut. The whole assembly works a bit like a carjack used to lift a vehicle: For example when changing a tire. You swivel, and the jackscrew rotates, pulling the so-called acme nuts inward and pushing the car up (see Fig. 2.2).

In the MD-80 trim system, the front part of the horizontal stabilizer is connected to a nut that drives up and down a vertical jackscrew. An electrical trim motor rotates the jackscrew, which in turn drives the nut up or down. The nut then pushes the whole horizontal tail up or down. Adequate lubrication is critical for the functioning of a jackscrew and nut assembly. Without enough grease, the constant grinding will wear out the thread on either the nut or the screw (in this case the screw is deliberately made of harder material, wearing the nut out first). The thread actually carries the entire load that is imposed on the vertical tail during flight. This is a load of around 5000 pounds, similar to the weight of a whole family van hanging by the thread of a jackscrew and nut assembly. Were the thread to wear out on an MD-80, the nut would fail to catch the threads of the jackscrew. Aerodynamic forces then push the horizontal tailplane (and the nut) to its stop way out of the normal range, rendering the aircraft uncontrollable in the pitch axis, which is essentially what happened to Alaska 261. Even the stop failed because of the pressure. A so-called torque tube runs through the

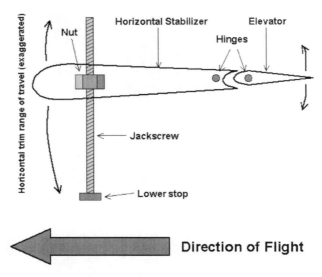

FIG. 2.2.   Simplified working of the trim mechanism in the tail of an MD-80-type aircraft. The horizontal stabilizer is hinged at the back and connected to the jackscrew through the nut. The stabilizer is tilted up and down by rotation of the jackscrew.

jackscrew in order to provide redundancy (instead of having two jack-screws, like in the preceding DC-8 model). But even the torque tube failed in Alaska 261.

None of this is supposed to happen, of course. When it first launched the aircraft in the mid 1960s, Douglas recommended that operators lubricate the trim jackscrew assembly every 300 to 350 flight hours. For typical commercial usage, that could mean grounding the airplane for such maintenance every few weeks. Immediately, the sociotechnical, organizational systems surrounding the operation of the technology began to adapt, and set the system on its course to drift. Through a variety of changes and developments in maintenance guidance for the DC-9/MD-80 series aircraft, the lubrication interval was extended. As we see later, these extensions were hardly the product of manufacturer recommendations alone, if at all. A much more complex and constantly evolving web of committees with representatives from regulators, manufacturers, subcontractors, and operators was at the heart of a fragmented, discontinuous development of maintenance standards, documents, and specifications. Rationality for maintenance-interval decisions was produced relatively locally, relying on incomplete, emerging information about what was, for all its deceiving basicness, still uncertain technology. Although each decision was locally rational, making sense for decision makers in their time and place, the global picture became one of drift toward disaster, significant drift.

Starting from a lubrication interval of 300 hours, the interval at the time of the Alaska 261 accident had moved up to 2,550 hours, almost an order of magnitude more. As is typical in the drift toward failure, this distance was not bridged in one leap. The slide was incremental: step by step, decision by decision. In 1985, jackscrew lubrication was to be accomplished every 700 hours, at every other so-called maintenance *B check* (which occurs every 350 flight hours). In 1987, the B-check interval itself was increased to 500 flight hours, pushing lubrication intervals to 1,000 hours. In 1988, B checks were eliminated altogether, and tasks to be accomplished were redistributed over A and C checks. The jackscrew assembly lubrication was to be done each eighth 125-hour A check: still every 1,000 flight hours. But in 1991, A-check intervals were extended to 150 flight hours, leaving a lubrication every 1200 hours. Three years later, the A-check interval was extended again, this time to 200 hours. Lubrication would now happen every 1,600 flight hours. In 1996, the jackscrew-assembly lubrication task was removed from the A check and moved instead to a so-called task card that specified lubrication every 8 months. There was no longer an accompanying flight-hour limit. For Alaska Airlines, 8 months translated to about 2,550 flight hours. The jackscrew recovered from the ocean floor, however, revealed no evidence that there had been adequate lubrication at the previous interval at all. It might have been more than 5,000 hours since it had last received a coat of fresh grease (see Fig. 2.3).

With as much lubrication as it originally recommended, Douglas thought it had no reason to worry about thread wear. So before 1967, the manufac-

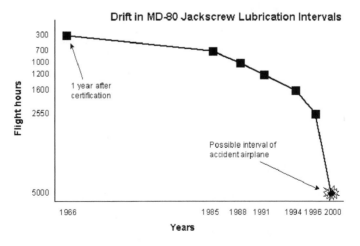

FIG. 2.3.   Drifting into failure over the decades: The jackscrew lubrication interval gradually got extended (almost by a factor of 10) until the Alaska Airlines 261 accident.

turer provided or recommended no check of the wear of the jackscrew assembly. The trim system was supposed to accumulate 30,000 flight hours before it would need replacement. But operational experience revealed a different picture. After only a year of DC-9 flying, Douglas received reports of thread wear significantly in excess of what had been predicted. In response, the manufacturer recommended that operators perform a so-called end-play check on the jackscrew assembly at every maintenance C check, or every 3,600 flight hours. The end-play check uses a restraining fixture that puts pressure on the jackscrew assembly, simulating the aerodynamic load during normal flight. The amount of play between nut and screw, gauged in thousandths of an inch, can then be read off an instrument. The play is a direct measure of the amount of thread wear.

From 1985 onward, end-play checks at Alaska became subject to the same kind of drift as the lubrication intervals. In 1985, end-play checks were scheduled every other C check, as the required C checks consistently came in around 2,500 hours, which was rather ahead of the recommended 3,600 flight hours, unnecessarily grounding aircraft. By scheduling an end-play test every other C check, though, the interval was extended to 5,000 hours. By 1988, C-check intervals themselves were extended to 13 months, with no accompanying flight-hour limit. End-play checks were now performed every 26 months, or about every 6,400 flight hours. In 1996, C-check intervals were extended once again, this time to 15 months. This stretched the flight hours between end-play tests to about 9,550. The last end-play check of the accident airplane was conducted at the airline maintenance facility in Oakland, California in 1997. At that time, play between nut and screw was found to be exactly at the allowable limit of .040 inches. This introduced considerable uncertainty. With play at the allowable limit, what to do? Release the airplane and replace parts the next time, or replace the parts now? The rules were not clear. The so-called AOL 9-48A said that "jackscrew assemblies could remain in service as long as the end-play measurement remained within the tolerances (between 0.003 and 0.040 inch)" (National Transportation Safety Board, or NTSB, 2002; p. 29). It was still 0.040 inches, so the aircraft could technically remain in service. Or could it? How quickly would the thread wear from there on? Six days, several shift changes and another, more favorable end-play check later, the airplane was released. No parts were replaced: They were not even in stock in Oakland. The airplane "departed 0300 local time. So far so good," the graveyard shift turnover plan noted (NTSB, 2002, p. 53). Three years later, the trim system snapped and the aircraft disappeared into the ocean not far away. Between 2,500 and 9,550 hours there had been more drift toward failure (see Fig. 2.4). Again, each extension made local sense, and was only an increment away from the previously established norm. No rules were violated, no laws broken. Even the regulator concurred with the changes in end-play check in-

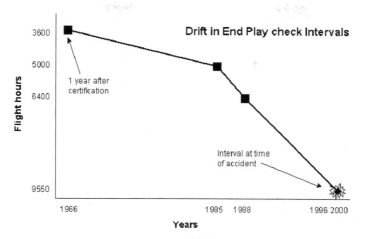

FIG. 2.4.   More drift into failure: The end play check interval (which gauges thread wear on the jackscrew-nut assembly) was stretched from 3,600 to 9,550 flight hours.

tervals. These were normal people doing normal work around seemingly normal, stable technology.

The figures of drift into failure are easy to draw in hindsight. They are fascinating to look at, too. The realities they represent, however, were not similarly compelling to those on the inside of the system at the time. Why would these numbers, this numeric degeneration of double checking and servicing, be noteworthy? As an indication, MD-80 maintenance technicians were never required to record or keep track of the end play on the trim systems they measured. Even the manufacturer had expressed no interest in seeing these numbers or the slow, steady degeneration they may have revealed. If there was drift, in other words, no institutional or organizational memory would know it.

The pictures of drift reveal what. But they shed no light on why. Indeed, the greatest conundrum since Turner (1978) has to been to elucidate why the slide into disaster, so easy to see and depict in retrospect, is missed by those who inflict it on themselves. Judging, after the fact, that there was a failure of foresight is easy: All you need to do is plot the numbers and spot the slide into disaster. Standing amid the rubble, it is easy to marvel at how misguided or misinformed people must have been. But why is it that the conditions conducive to an accident were never acknowledged or acted on by those on the inside of the system—those whose job it was to not have such accidents happen? Foresight is not hindsight. There is a profound revision of insight that turns on the present. It converts a once vague, unlikely future into an immediate, certain past. The future, said David Woods (2003), seems implausible before an accident ("No, that won't happen to us"). But after an accident, the past seems incredible ("How could we not have seen that this was going to happen to us!"). What is now seen as extraordinary was once ordinary. The decisions, trade-offs, preferences and priorities that seem so out of the ordinary and immoral after an accident, were once normal and common sensical to those who contributed to its incubation.

BANALITY, CONFLICT, AND INCREMENTALISM

Sociological research (e.g., Perrow, 1984; Snook, 2000; Vaughan, 1996; Weick, 1995), as well as prescient human factors work (Rasmussen & Svedung, 2000) and research on system safety (Leveson, 2002), has begun to sketch the contours of answers to the why of drift. Though different in background, pedigree, and much substantive detail, these works converge on important commonalities about the drift into failure. The first is that accidents, and the drift that precedes them, are associated with normal people doing normal work in normal organizations—not with miscreants engaging in immoral deviance. We can call this the *banality-of-accidents* thesis. Second, most works have at their heart a conflictual model: Organizations that involve safety-critical work are essentially trying to reconcile irreconcilable goals (staying safe and staying in business). Third, drifting into failure is incremental. Accidents do not happen suddenly, nor are they preceded by monumentally bad decisions or bizarrely huge steps away from the ruling norm.

The banality-of-accidents thesis says that the potential for having an accident grows as a normal by-product of doing normal business under normal pressures of resource scarcity and competition. No system is immune to the pressures of scarcity and competition, well, almost none. The only transportation system that ever approximated working in a resource-unlimited universe was NASA during the early Apollo years (a man had to be put on the moon, whatever the cost). There was plenty of money, and plenty of highly motivated talent. But even here technology was uncertain, faults and failures not uncommon, and budgetary constraints got imposed quickly and increasingly tightly. Human resources and talent started to drain away. Indeed, even such noncommercial enterprises know resource scarcity: Government agencies such as NASA or safety regulators may lack adequate financing, personnel or capacity to do what they need to do. With respect to the Alaska 261 accident, for example, a new regulatory inspection program, called the Air Transporation Oversight System (ATOS), was put into use in 1998 (2 years prior). It drastically reduced the amount of time inspectors had for actual surveillance activities. A 1999 memo by a regulator field-office supervisor in Seattle offered some insight:

We are not able to properly meet the workload demands. Alaska Airlines has expressed continued concern over our inability to serve it in a timely manner. Some program approvals have been delayed or accomplished in a rushed manner at the "eleventh hour," and we anticipate this problem will intensify with time. Also, many enforcement investigations . . . have been delayed as a result of resource shortages. [If the regulator] continues to operate with the existing limited number of airworthiness inspectors . . . diminished surveillance is imminent and the risk of incidents or accidents at Alaska Airlines is heightened. (NTSB, 2002, p. 175)

Adapting to resource pressure, approvals were delayed or rushed, surveillance was reduced. Yet doing business under pressures of resource scarcity is normal: Scarcity and competition are part and parcel even of doing inspection work. Few regulators anywhere will ever claim that they have adequate time and personnel resources to carry out their mandates. Yet the fact that resource pressure is normal does not mean that it has no consequences. Of course the pressure finds ways out. Supervisors write memos, for example. Battles over resources are fought. Trade-offs are made. The pressure expresses itself in the common organizational, political wrangles over resources and primacy, in managerial preferences for certain activities and investments over others, and in almost all engineering and operational trade-offs between strength and cost, between efficiency and diligence. In fact, working successfully under pressures and resource constraints is a source of professional pride: Building something that is strong and light, for example, marks the expert in the aeronautical engineer. Procuring and nursing into existence a system that has both low development costs and low operational costs (these are typically each other's inverse) is the dream of most investors and many a manager. Being able to create a program that putatively allows better inspections with fewer inspectors may win a civil servant compliments and chances at promotion, while the negative side effects of the program are felt primarily in some far-away field office.

Yet the major engine of drift hides somewhere in this conflict, in this tension between operating safely and operating at all, between building safely and building at all. This tension provides the energy behind the slow, steady disengagement of practice from earlier established norms or design constraints. This disengagement can eventually become drift into failure. As a system is taken into use, it learns, and as it learns, it adapts:

Experience generates information that enables people to fine-tune their work: fine-tuning compensates for discovered problems and dangers, removes redundancy, eliminates unnecessary expense, and expands capacities. Experience often enables people to operate a sociotechnical system for much lower cost or to obtain much greater output than the initial design assumed. (Starbuck & Milliken, 1988, p. 333)

This fine-tuning drift toward operational safety margins is one testimony to the limits of the structuralist systems safety vocabulary in vogue today. We think of safety cultures as learning cultures: cultures that are oriented toward learning from events and incidents. But learning cultures are neither unique (because every open system in a dynamic environment necessarily learns and adapts) nor necessarily positive: Starbuck and Milliken highlighted how an organization can learn to "safely" borrow from safety while achieving gains in other areas. Drift into failure could not happen without learning. Following this logic, systems that are bad at learning and bad at adapting may well be less likely to drift into failure.

A critical ingredient of this learning is the apparent insensitivity to mounting evidence that, from the position of retrospective outsider, could have shown how bad the judgments and decisions actually are. This is how it looks from the position of retrospective outsider: The retrospective outsider sees a failure of foresight. From the inside, however, the abnormal is pretty normal, and making trade-offs in the direction of greater efficiency is nothing unusual. In making these trade-offs, however, there is a feedback imbalance. Information on whether a decision is costeffective or efficient can be relatively easy to get. An early arrival time is measurable and has immediate, tangible benefits. How much is or was borrowed from safety in order to achieve that goal, however, is much more difficult to quantify and compare. If it was followed by a safe landing, apparently it must have been a safe decision. Extending a lubrication interval similarly saves immediately measurable time and money, while borrowing from the future of an apparently problem-free jackscrew assembly. Each consecutive empirical success (the early arrival time is still a safe landing; the jackscrew assembly is still operational) seems to confirm that fine-tuning is working well: The system can operate equally safely, yet more efficiently. As Weick (1993) pointed out, however, safety in those cases may not at all be the result of the decisions that were or were not made, but rather an underlying stochastic variation that hinges on a host of other factors, many not easily within the control of those who engage in the fine-tuning process. Empirical success, in other words, is not proof of safety. Past success does not guarantee future safety. Borrowing more and more from safety may go well for a while, but you never know when you are going to hit. This moved Langewiesche (1998) to say that Murphy's law is wrong: Everything that can go wrong usually goes right, and then we draw the wrong conclusion.

The nature of this dynamic, this fine-tuning, this adaptation, is incremental. The organizational decisions that are seen as "bad decisions" after the accident (even though they seemed like perfectly good ideas at the time) are seldom big, risky, order-of-magnitude steps. Rather, there is a succession of increasingly bad decisions, a long and steady progression of small, incremental steps that unwittingly take an organization toward disaster. Each step away from the original norm that meets with empirical success (and no obvious sacrifice of safety) is used as the next basis from which to depart just that little bit more again. It is this incrementalism that makes distinguishing the abnormal from the normal so difficult. If the difference between what "should be done" (or what was done successfully yesterday) and what is done successfully today is minute, then this slight departure from an earlier established norm is not worth remarking or reporting on. Incrementalism is about continued normalization: It allows normalization and rationalizes it.

Drift Into Failure and Incident Reporting

Can incident reporting not reveal a drift into failure? This would seem to be a natural role of incident reporting, but it is not so easy. The normalization that accompanies drift into failure (an end-play check every 9,550 hours is "normal," even approved by the regulator, no matter that the original interval was 2,500 hours) severely challenges the ability of insiders to define incidents. What is an incident? Before 1985, failing to perform an end-play check every 2,500 hours could be considered an incident, and supposing the organization had a means for reporting it, it may even have been considered as such. But by 1996, the same deviance was normal, regulated even. By 1996, the same failure was no longer an incident. And there was much more. Why report that lubricating the jackscrew assembly often has to be done at night, in the dark, outside the hanger, standing in the little basket of a lift truck at a soaring height above the ground, even in the rain? Why report that you, as a maintenance mechanic have to fumble your way through two tiny access panels that hardly allow room for one human hand—let alone space for eyes to see what is going on inside and what needs to be lubricated—if that is what you have to do all the time? In maintenance, this is normal work, it is the type of activity required to get the job done. The mechanic responsible for the last lubrication of the accident airplane told investigators that he had taken to wearing a battery-operated head lamp during night lubrication tasks, so that he had his hands free and could see at least something. These things are normal, they are not reportworthy. They do not qualify as incidents. Why report that the end-play checks are performed with one restraining fixture (the only one in the entire airline, fabricated in-house, nowhere near the manufacturer's specifications), if that is what you use every time you do an end-play check? Why report that end-play checks, either on the airplane or on the bench, generate widely varying measures, if that is what they do all the time, and if that is what maintenance work is often about? It is normal, it is not an incident. Even if the airline had had a reporting culture, even if it had had a learning culture, even if it had had a just culture so that people would feel secure in sending in their reports without fear of retribution, these would not be incidents that would turn up in the system. This is the banality of accidents thesis. These are not incidents. In 10^{-7} systems, incidents do not precede accidents. Normal work does. In these systems:

> accidents are different in nature from those occurring in safe systems: in this case accidents usually occur in the absence of any serious breakdown or even of any serious error. They result from a combination of factors, none of which can alone cause an accident, or even a serious incident; therefore these combinations remain difficult to detect and to recover using traditional safety analysis logic. For the same reason, reporting becomes less relevant in predicting major disasters. (Amalberti, 2001, p. 112)

Even if we were to direct greater analytic force onto our incident-reporting databases, this may still not yield any predictive value for accidents beyond 10^{-7}, simply because the data is not there. The databases do not contain, in any visible format, the ingredients of accidents that happen beyond 10^{-7}. Learning from incidents to prevent accidents beyond 10^{-7} may well be impossible. Incidents are about independent failures and errors, noticed and noticeable by people on the inside. But these independent errors and failures no longer make an appearance in the accidents that happen beyond 10^{-7}. The failure to adequately see the part to be lubricated (that nonredundant, single-point, safety-critical part), the failure to adequately and reliably perform an end-play check—none of this appears in incident reports. But it is deemed "causal" or "contributory" in the accident report. The etiology of accidents in 10^{-7} systems, then, may well be fundamentally different from that of incidents, hidden instead in the residual risks of doing normal business under normal pressures of scarcity and competition. This means that the so-called common-cause hypothesis (which holds that accidents and incidents have common causes and that incidents are qualitatively identical to accidents except for being just one step short) is probably wrong at 10^{-7} and beyond:

> ... Reports from accidents such as Bhopal, Flixborough, Zeebrugge and Chernobyl demonstrate that they have not been caused by a coincidence of independent failures and human errors. They were the effect of a systematic migration of organizational behavior toward accident under the influence of pressure toward cost-effectiveness in an aggressive, competitive environment. (Rasmussen & Svedung, 2000, p. 14)

Despite this insight, independent errors and failures are still the major return of any accident investigation today. The 2002 NTSB report, following Newtonian–Cartesian logic, spoke of deficiencies in Alaska Airlines' maintenance program, of shortcomings in regulatory oversight, of responsibilities not fulfilled, of flaws and failures and breakdowns. Of course, in hindsight they may well be just that. And finding faults and failures is fine because it gives the system something to fix. But why did nobody at the time see these oh-so apparent faults and failures for what they (in hindsight) were? This is where the structuralist vocabulary of traditional human factors and systems safety is most limited, and limiting. The holes found in the layers of defense (the regulator, the manufacturer, the operator, the maintenance facility and, finally, the technician) are easy to discover once the rubble is strewn before one's feet. Indeed, one common critique of structuralist models is that they are good at identifying deficiencies, or latent failures, post-mortem. Yet these deficiencies and failures are not seen as such, nor easy to see as such, by those on the inside (or even those relatively on the outside, like the regulator!) before the accident happens. Indeed, structuralist models can capture the deficiencies that result from drift very well: They accurately identify latent failures or resident pathogens in organizations and can locate holes in layers of defense. But the build-up of latent failures, if that is what you want to call them, is not modeled. The process of erosion, of attrition of safety norms, of drift toward margins, cannot be captured well by structuralist approaches, for those are inherently metaphors for resulting forms, not models oriented at processes of formation. Structuralist models are static.

Although the structuralist models of the 1990s are often called "system models" or "systemic models," they are a far cry from what is actually considered systems thinking (e.g., Capra, 1982). The systems part of structuralist models has so far largely been limited to identifying, and providing a vocabulary for, the upstream structures (blunt ends) behind the production of errors at the sharp end. The systems part of these models is a reminder that there is context, that we cannot understand errors without going into the organizational background from which they hail. All of this is necessary, of course, as errors are still all too often taken as the legitimate conclusion of an investigation (just look at the spoiler case with "breakdown in CRM" as cause). But reminding people of context is no substitute for beginning to explain the dynamics, the subtle, incremental processes that lead to, and normalize, the behavior eventually observed. This requires a different perspective for looking at the messy interior of organizations, and a different language to cast the observations in. It requires human factors and system safety to look for ways that move toward real systems thinking, where accidents are seen as an emergent feature of organic, ecological, transactive processes, rather than just the end-point of a trajectory through holes in layers of defense. Structuralist approaches, and fixing the things they point us to, may not help much in making further progress on safety:

We should be extremely sensitive to the limitations of known remedies. While good management and organizational design may reduce accidents in certain systems, they can never prevent them . . . The causal mechanisms in this case suggest that technical system failures may be more difficult to avoid than even the most pessimistic among us would have believed. The effect of unacknowledged and invisible social forces on information, interpretation, knowledge, and—ultimately—action, are very difficult to identify and to control. (Vaughan, 1996, p. 416)

Yet the retrospective explanatory power of structuralist models makes them the instruments of choice for those in charge of managing safety. Indeed, the idea of a banality of accidents has not always easily found traction outside academic circles. For one thing, it is scary. It makes the potential for failure commonplace, or relentlessly inevitable (Vaughan, 1996). This can make accident models practically useless and managerially demoralizing. If the potential for failure is everywhere, in everything we do, then why try to avoid it? If an accident has no causes in the traditional sense, then why try to fix anything? Such questions are indeed nihilist, fatalist. It is not surprising, then, that resistance against the possible world lurking behind their answers takes many forms. Pragmatic concerns are directed toward control, toward hunting down the broken parts, the bad guys, the violators, the incompetent mechanics. Why did this one technician not perform the last lubrication of the accident airplane jackscrew as he should have? Pragmatic concerns are about finding the flaws, identifying the weak areas and trouble spots, and fixing them before they cause real problems. But those pragmatic concerns find neither a sympathetic ear nor a constructive lexicon in the misers about drift into failure, for drift into failure is hard to spot, certainly from the inside.

TOWARD SYSTEMS THINKING

If we want to understand failures beyond 10^{-7}, we have to stop looking for failures. It are no longer failures that go into creating these failures—it is normal work. Thus the banality of accidents makes their study philosophically philistine. It shifts the object of examination away from the darker sides of humanity and unethical corporate malgovernance, and toward pedestrian, everyday decisions of normal, everyday people under the influence of normal, everyday pressures. The study of accidents is rendered dramatic or fascinating only because of the potential outcome, not because of the processes that incubate it (which in itself can be fascinating, of course). Having studied the Challenger Space Shuttle disaster extensively, Diane Vaughan (1996) was forced to conclude that this type of accident is not caused by a series of component failures, even if component failures are the result. Instead, together with other sociologists, she pointed to an indigenousness of mistake, to mistakes and breakdown as systematic, normal byproducts of an organization's work processes:

> Mistake, mishap, and disaster are socially organized and systematically produced by social structures. No extraordinary actions by individuals explain what happened: no intentional managerial wrongdoing, no rule violations, no conspiracy. These are mistakes embedded in the banality of organizational life and facilitated by environments of scarcity and competition, uncertain technology, incrementalism, patterns of information, routinization and organizational structures. (p. xiv)

If we want to understand, and become able to prevent, failure beyond 10^{-7}, this is where we need to look. Forget wrongdoing. Forget rule violations. Forget errors. Safety, and the lack of it, is an emergent property. What we need to study instead is patterns of information, the uncertainties in operating complex technology and the ever-evolving and imperfect sociotechnical systems surrounding it to make that operation happen, the influence of scarcity and competition on those systems, and how they set in motion an incrementalism (itself an expression of organizational learning or adaptation under those pressures). To understand safety, an organization needs to capture the dynamics in the banality of its organizational life and begin to see how the emergent collective moves toward the boundaries of safe performance.

Systems as Dynamic Relationships

Capturing and describing the processes by which organizations drift into failure requires systems thinking. Systems thinking is about relationships and integration. It sees a sociotechnical system not as a structure consisting of constituent departments, blunt ends and sharp ends, deficiencies and flaws, but as a complex web of dynamic, evolving relationships and transactions. Instead of building blocks, the systems approach emphasizes principles of organization. Understanding the whole is quite different from understanding an assembly of separate components. Instead of mechanical linkages between components (with a cause and an effect), it sees transactions—simultaneous and mutually interdependent interactions. Such emergent properties are destroyed when the system is dissected and studied as a bunch of isolated components (a manager, department, regulator, manufacturer, operator). Emergent properties do not exist at lower levels; they cannot even be described meaningfully with languages appropriate for those lower levels.

Take the lengthy, multiple processes by which maintenance guidance was produced for the DC-9 and later the MD-80 series aircraft. Separate components (such as regulator, manufacturer, operator) are difficult to distinguish, and the interesting behavior, the kind of behavior that helps drive drift into failure, emerges only as a result of complex relationships and transactions. At first thought, the creation of maintenance guidance would seem a solved problem. You build a product, you get the regulator to certify it as safe to use, and then you tell the user how to maintain it in order to keep it safe. Even the second step (getting it certified as safe) is nowhere near a solved problem, and is deeply intertwined with the third. More about that later: First the maintenance guidance. Alaska 261 reveals a large gap between the production of a system and its operation. Inklings of the gap appeared in observations of jackscrew wear that was higher than what the manufacturer expected. Not long after the certification of the DC-9, people began work to try to bridge the gap. Assembling people from across the industry, a Maintenance Guidance Steering Group (MSG) was set up to develop guidance documentation for maintaining large transport aircraft (NTSB, 2002), particularly the Boeing 747. Using this experience, another MSG developed a new guidance document in 1970, called MSG-2 (NTSB, 2002), which was intended to present a means for developing a maintenance program acceptable to the regulator, the operator, and the manufacturer. The many discussions, negotiations, and interorganizational collaborations underlying the development of an "acceptable maintenance program" showed that how to maintain a once certified piece of complex technology was not at all a solved problem. In fact, it was very much an emerging property: Technology proved less certain than it had seemed on the drawing board (e.g., the DC-9 jackscrew-wear rates were higher than predicted), and it was not before it hit the field of practice that deficiencies became apparent, if one knew where to look.

In 1980, through combined efforts of the regulator, trade and industry groups and manufacturers of both aircraft and engines in the United States as well as Europe, a third guidance document was produced, called MSG-3 (NTSB, 2002). This document had to deconfound earlier confusions, for example, between "hard-time" maintenance, "on-condition" maintenance, "condition-monitoring" maintenance, and "overhaul" maintenance. Revisions to MSG-3 were issued in 1988 and 1993. The MSG guidance documents and their revisions were accepted by the regulators, and used by so-called Maintenance Review Boards (MRB) that convene to develop guidance for specific aircraft models.

A Maintenance Review Board, or MRB, does not write guidance itself, however; this is done by industry steering committees, often headed by a regulator. These committees in turn direct various working groups. Through all of this, so-called on-aircraft maintenance planning (OAMP) documents get produced, as well as generic task cards that outline specific maintenance jobs. Both the lubrication interval and the end-play check for MD-80 trim jackscrews were the constantly changing products of these evolving webs of relationships between manufacturers, regulators, trade groups, and operators, who were operating off of continuously renewed operational experience, and a perpetually incomplete knowledge base about the still uncertain technology (remember, end-play test results were not recorded or tracked). So what are the rules? What should the standards be? The introduction of a new piece of technology is followed by negotiation, by discovery, by the creation of new relationships and rationalities. "Technical systems turn into models for themselves," said Weingart (1991), "the observation of their functioning, and especially their malfunctioning, on a real scale is required as a basis for further technical development" (p. 8). Rules and standards do not exist as unequivocal, aboriginal markers against a tide of incoming operational data (and if they do, they are quickly proven useless or out of date). Rather, rules and standards are the constantly updated products of the processes of conciliation, of give and take, of the detection and rationalization of new data. As Brian Wynne (1988) said:

> Beneath a public image of rule-following behavior and the associated belief that accidents are due to deviation from those clear rules, experts are operating with far greater levels of ambiguity, needing to make expert judgments in less than clearly structured situations. The key point is that their judgments are not normally of a kind—how do we design, operate and maintain the system according to "the" rules? Practices do not follow rules, rather, rules follow evolving practices. (p. 153)

Setting up the various teams, working groups, and committees was a way of bridging the gap between building and maintaining a system, between producing it and operating it. Bridging the gap is about adaptation—adaptation to newly emerging data (e.g., surprising wear rates) about an uncertain technology. But adaptation can mean drift. And drift can mean breakdown.

MODELING LIVE SOCIOTECHNICAL SYSTEMS

What kind of safety model could capture such adaptation, and predict its eventual collapse? Structuralist models are limited. Of course, we could claim that the lengthy lubrication interval and the unreliable end-play check were structural deficiencies. Were they holes in layers of defense? Absolutely. But such metaphors do not help us look for where the hole occurred, or why. There is something complexly organic about MSGs, something ecological, that is lost when we model them as a layer of defense with a hole in it; when we see them as a mere deficiency or a latent failure. When we see systems instead as internally plastic, as flexible, as organic, their functioning is controlled by dynamic relations and ecological adaptation, rather than by rigid mechanical structures. They also exhibit self-organization (from year to year, the makeup of MSGs was different) in response to environmental changes, and self-transcendence: the ability to reach out beyond currently known boundaries and learn, develop and perhaps improve. What is needed is not yet another structural account of the end result of organizational deficiency. What is needed instead is a more functional account of living processes that coevolve with respect to a set of environmental conditions, and that maintain a dynamic and reciprocal relation with those conditions (see Heft, 2001). Such accounts need to capture what happens within an organization, with the gathering of knowledge and creation of rationality within workgroups, once a technology gets fielded. A functional account could cover the organic organization of maintenance steering groups and committees, whose makeup, focus, problem definition, and understanding coevolved with emerging anomalies and growing knowledge about an uncertain technology.

A model that is sensitive to the creation of deficiencies, not just to their eventual presence, makes a sociotechnical system come alive. It must be a model of processes, not just one of structure. Extending a lineage of cybernetic and systems-engineering research, Nancy Leveson (2002) proposed that control models can fulfill part of this task. Control models use the ideas of hierarchies and constraints to represent the emergent interactions of a complex system. In their conceptualization, a sociotechnical system consists of different levels, where each superordinate level imposes constraints on (or controls what is going on in) subordinate levels. Control models are one way to begin to map the dynamic relationships between different levels within a system—a critical ingredient of moving toward true systems thinking (where dynamic relationships and transactions are dominant, not structure and components). Emergent behavior is associated with the limits or constraints on the degrees of freedom of a particular level.

The division into hierarchical levels is an analytic artifact necessary to see how system behavior can emerge from those interactions and relationships. The resulting levels in a control model are of course a product of the analyst who maps the model onto the sociotechnical system. Rather than reflections of some reality out there, the patterns are constructions of a human mind looking for answers to particular questions. For example, a particular MSG would probably not see how it is superordinate to some level and imposing constraints on it, or subordinate to some other and thus subject to its constraints. In fact, a one-dimensional hierarchical representation (with only up and down along one direction) probably oversimplifies the dynamic web of relationships surrounding (and determining the functioning of) any such multiparty, evolving group as an MSG. But all models are simplifications, and the levels analogy can be helpful for an analyst who has particular questions in mind (e.g., why did these people at this level or in this group make the decisions they did, and why did they see that as the only rational way to go?).

Control among levels in a sociotechnical system is hardly ever perfect. In order to control effectively, any controller needs a good model of what it is supposed to control, and it requires feedback about the effectiveness of its control. But such internal models of the controllers easily become inconsistent with, and no longer match, the system to be controlled (Leveson, 2002). Buggy control models are true especially with uncertain, emerging technology (including trim jackscrews) and the maintenance requirements surrounding them. Feedback about the effectiveness of control is incomplete and can be unreliable too. A lack of jackscrew-related incidents may provide the illusion that maintenance control is effective and that intervals can be extended, whereas the paucity of risk actually depends on factors quite outside the controller's scope. In this sense, the imposition of constraints on the degrees of freedom is mutual between levels and not just top down: If subordinate levels generate imperfect feedback about their functioning, then higher order levels do not have adequate resources (degrees of freedom) to act as would be necessary. Thus the subordinate level imposes constraints on the superordinate level by not telling (or not being able to tell) what is really going on. Such a dynamic has been noted in various cases of drift into failure, including the Challenger Space Shuttle disaster (see Feynman, 1988).

Drift Into Failure as Erosion of Constraints and Eventual Loss of Control

Nested control loops can make a model of a sociotechnical system come alive more easily than a line of layers of defense. In order to model drift, it has to come alive. Control theory sees drift into failure as a gradual erosion of the quality or the enforcement of safety constraints on the behavior of subordinate levels. Drift results from either missing or inadequate constraints on what goes on at other levels. Modeling an accident as a sequence of events, in contrast, is really only modeling the end product of such erosion and loss of control. If safety is seen as a control problem, then events (just like the holes in layers of defense) are the results of control problems, not the causes that drive a system into disaster. A sequence of events, in other words, is at best the starting point of modeling an accident, not the analytic conclusion. The processes that generate these weaknesses are in need of a model.

One type of erosion of control occurs because original engineering constraints (e.g., 300-hour intervals) are loosened in response to the accumulation of operational experience. A variety of Starbuck and Milliken's (1988) "fine-tuning," in other words. This does not mean that the kind of ecological adaptation in system control is fully rational, or that it makes sense even from a global perspective on the overall evolution and eventual survival of the system. It does not. Adaptations occur, adjustments get made, and constraints get loosened in response to local concerns with limited time-horizons. They are all based on uncertain, incomplete knowledge. Often it is not even clear to insiders that constraints have become less tight as a result of their decisions in the first place, or that it matters if it is. And even when it is clear, the consequences may be hard to foresee, and judged to be a small potential loss in relation to the immediate gains. As Leveson (2002) put it, experts do their best to meet local conditions, and in the busy daily flow and complexity of activities they may be unaware of any potentially dangerous side effects of those decisions. It is only with the benefit of hindsight or omniscient oversight (which is utopian) that these side effects can be linked to actual risk. Jensen (1996) describes it as such:

> We should not expect the experts to intervene, nor should we believe that they always know what they are doing. Often they have no idea, having been blinded to the situation in which they are involved. These days, it is not unusual for engineers and scientists working within systems to be so specialized that they have long given up trying to understand the system as a whole, with all its technical, political, financial, and social aspects. (p. 368)

Being a member of a system, then, can make systems thinking all but impossible. Perrow (1984) made this argument very persuasively, and not just for the system's insiders. An increase in system complexity diminishes the system's transparency: Diverse elements interact in a greater variety of ways that are difficult to foresee, detect, or even comprehend. Influences from outside the technical knowledge base (those "political, financial, and social aspects" of Jensen, 1996, p. 368) exert a subtle but powerful pressure on the decisions and trade-offs that people make, and constrain what is seen as a rational decision or course of action at the time (Vaughan, 1996). Thus, even though experts may be well-educated and motivated, a "warning of an incomprehensible and unimaginable event cannot be seen, because it cannot be believed" (Perrow, 1984, p. 23). How can experts and other decision makers inside organizational systems make sense of the available indicators of system safety performance? Making sure that experts and other decision makers are well informed is in itself an empty pursuit. What well informed really means in a complex organizational setting is infinitely negotiable, and clear criteria for what constitutes enough information are impossible to obtain. As a result, the effect of beliefs and premises on decision making and the creation of rationality can be considerable. Weick (1995, p. 87) pointed out that "seeing what one believes and not seeing that for which one has no beliefs are central to sensemaking. Warnings of the unbelievable go unheeded." That which cannot be believed will not be seen. This confirms the earlier pessimism about the value of incident reporting beyond 10^{-7}. Even if relevant events and warnings end up in the reporting system (which is doubtful because they are not seen as warnings even by those who would do the reporting), it is even more generous to presume that further expert analysis of such incident databases could succeed in coaxing the warnings into view.

The difference, then, between expert insight at the time and hindsight (after an accident) is tremendous. With hindsight, the internal workings of the system may become lucid: The interactions and side effects are rendered visible. And with hindsight, people know what to look for, where to dig around for the rot, the missing connections. Triggered by the Alaska 261 accident, the regulator launched a special probe into the maintenance-control system at Alaska Airlines. It found that procedures in place at the company were not followed, that controls in place were clearly not effective, that authority and responsibility were not well defined, that control of the maintenance-deferral systems was missing, and that quality-control and quality-assurance programs and departments were ineffective. It also found incomplete C-check paperwork, discrepancies of shelf-life expiration dates of parts, a lack of engineering approval of maintenance work-card modifications and inadequate tool calibrations. Maintenance manuals did not specify procedures or objectives for on-the-job training of mechanics, and key management positions (e.g., safety) were not filled or did not exist. Indeed, constraints imposed on other organizational levels were nonexistent, dysfunctional, or eroded.

But seeing holes and deficiencies in hindsight is not an explanation of the generation or continued existence of those deficiencies. It does not help predict or prevent failure. Instead, the processes by which such decisions come about, and by which decision makers create their local rationality, are one key to understanding how safety can erode on the inside of a complex, sociotechnical system. Why did these things make sense to organizational decision makers at the time? Why was it all normal, why was it not reportworthy, not even for the regulator tasked with overseeing these processes? The questions hang in the air. Little evidence is available from the (already huge) NTSB investigation on such interorganizational processes or how they produced a particular conceptualization of risk. The report, like others, is testimony to the structuralist, mechanistic tradition in accident probes to date, applied even to investigative forays into social-organizational territory.

The Creation of Local Rationality

The question is, how do insiders make those numerous little and larger trade-offs that together contribute to erosion, to drift? How is it that these seemingly harmless decisions can incrementally move a system to the edge of disaster? As indicated earlier, a critical aspect of this dynamic is that people in decision-making roles on the inside of a sociotechnical system miss or underestimate the global side effects of their locally rational decisions. As an example, the MSG-3 MD-80 MRB (if you just lost it there, do not worry, other people must have too) considered the 3,600 hour jackscrew lubrication task change as part of the larger C-check package (NTSB, 2002). The review board did not consult the manufacturer's design engineers, nor did it make them aware of the extension. The manufacturer's initial OAMP document for the DC-9 and MD-80 lubrication, specifying an already extended 600- to 900-hour interval (departing from the 1964 recommendation for 300 hours), was also not considered in MSG-3. From a local perspective, with the pressure of time limits and constraints on available knowledge, the decision to extend the interval without adequate expert input must have made sense. People consulted at the time must have been deemed adequate and sufficiently expert in order to feel comfortable enough to continue. The creation of rationality must have been seen as satisfactory. Otherwise, it is hard to believe that MSG-3 would have proceeded as it did. But the eventual side effects of these smaller decisions were not foreseen. From the larger perspective, the gap between production and operation, between making and maintaining a product, was once again allowed to widen. A relationship that had been instrumental in helping bridge that gap (consulting with the original design engineers who make the aircraft, to inform those who maintain it), a relationship from history to (then) present, was severed. A transaction was not completed.

If not foreseeing side effects made sense for MSG-3 MD-80 MRB (and this may well have been a banal result of the sheer complexity and paper load of the work mandated), it may make sense for participants in a next sociotechnical system too. These decisions are sound when set against local judgment criteria; given the time and budget pressures and short-term incentives that shape behavior. Given the knowledge, goals, and attentional focus of the decision makers and the nature of the data available to them at the time, it made sense. It is in these normal, day-to-day processes, where we can find the seeds of organizational failure and success. And it is these processes we must turn to in order to find leverage for making further progress on safety. As Rasmussen and Svedung (2000) put it:

> To plan for a proactive risk management strategy, we have to understand the mechanisms generating the actual behavior of decision-makers at all levels . . . an approach to proactive risk management involves the following analyses:

- A study of normal activities of the actors who are preparing the landscape of accidents during their normal work, together with an analysis of the work features that shape their decision making behavior
- A study of the present information environment of these actors and the information flow structure, analyzed from a control theoretic point of view. (p. 14)

Reconstructing or studying the "information environment" in which actual decisions are shaped, in which local rationality is constructed, can help us penetrate processes of organizational sensemaking. These processes lie at the root of organizational learning and adaptation, and thereby at the source of drift into failure. The two Space Shuttle accidents (Challenger in 1996 and Columbia in 2002) are highly instructive here, if anything because the Columbia Accident Investigation Board (CAIB), as well as later analyses of the Challenger disaster (e.g., Vaughan, 1996) represent significant (and, to date, rather unique) departures from the typical structuralist probes into such accidents. These analyses take normal organizational processes toward drift seriously, applying and even extending a language that helps us capture something essential about the continuous creation of local rationality by organizational decision makers.

One critical feature of the information environment in which NASA engineers made decisions about safety and risk was "bullets." Richard Feynman, who participated in the original Rogers Presidential Commission investigating the Challenger disaster, already fulminated against them and the way they collapsed engineering judgments into crack statements: "Then we learned about 'bullets'—little black circles in front of phrases that were supposed to summarize things. There was one after another of these little goddamn bullets in our briefing books and on the slides" (Feynman, 1988, p. 127).

Eerily, "bullets" appear again as an outcropping in the 2003 Columbia accident investigation. With the proliferation of commercial software for making "bulletized" presentations since Challenger, bullets proliferated as well. This too may have been the result of locally rational (though largely unreflective) trade-offs to increase efficiency: Bulletized presentations collapse data and conclusions and are dealt with more quickly than technical papers. But bullets filled up the information environment of NASA engineers and managers at the cost of other data and representations. They dominated technical discourse and, to an extent, dictated decision making, determining what would be considered as sufficient information for the issue at hand. Bulletized presentations were central in creating local rationality, and central in nudging that rationality ever further away from the actual risk brewing just below.

Edward Tufte (CAIB, 2003) analyzed one Columbia slide in particular, from a presentation given to NASA by a contractor in February 2003. The aim of the slide was to help NASA consider the potential damage to heat tiles created by ice debris that had fallen from the main fuel tank. (Damaged heat tiles triggered the destruction of Columbia on the way back into the earth's atmosphere, see Fig. 2.5.) The slide was used by the Debris Assessment Team in their presentation to the Mission Evaluation Room. It was entitled "Review of Test Data Indicates Conservatism for Tile Penetration," suggesting, in other words, that the damage done to the wing was not so bad (CAIB, 2003, p. 191). But actually, the title did not refer to predicted tile damage at all. Rather, it pointed to the choice of test models used to predict the damage. A more appropriate title, according to Tufte, would have been "Review of test data indicates irrelevance of two models." The reason was that the piece of ice debris that struck the Columbia was estimated to be 640 times larger than the data used to calibrate the model on which engineers based their damage assessments (later analysis showed that the debris object was actually 400 times larger). So the calibration models were not of much use: They hugely underestimating the actual impact of the debris. The slide went on to say that "significant energy" would be required to have debris from the main tank penetrate the (supposedly harder) tile coating of the Shuttle wing, yet that test results showed that this was possible at sufficient mass and velocity, and that, once the tiles were penetrated, significant damage would be caused. As Tufte observed, the vaguely quantitative word "significant" or "significantly" was used five times on the one slide, but its meaning ranged all the way from the ability to see it using those irrelevant calibration tests, through a difference of 640-fold, to

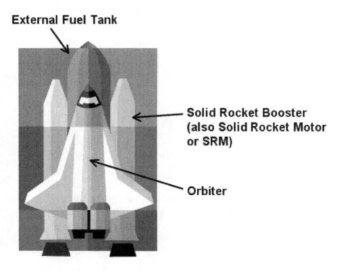

FIG. 2.5.   Location of the solid rocket boosters (Challenger) and external fuel tank (Columbia) on a Space Shuttle.

WHY DO SAFE SYSTEMS FAIL?   **41**

damage so great that everybody onboard would die. The same word, the same token on a slide, repeated five times, carried five profoundly (yes, significantly) different meanings, yet none of those were really made explicit because of the condensed format of the slide. Similarly, damage to the protective heat tiles was obscured behind one little word, *it*, in a sentence that read "Test results show that it is possible at sufficient mass and velocity" (CAIB, 2003, p. 191). The slide weakened important material, and the life-threatening nature of the data on it was lost behind bullets and abbreviated statements.

A decade and a half before, Feynman (1988) had discovered a similarly ambiguous slide about Challenger. In his case, the bullets had declared that the eroding seal in the field joints was "most critical" for flight safety, yet that "analysis of existing data indicates that it is safe to continue flying the existing design" (p. 137). The accident proved that it was not. Solid Rocket Boosters (or SRBs or SRMs) that help the Space Shuttle out of the earth's atmosphere are segmented, which makes ground transportation easier and has some other advantages. A problem that was discovered early in the Shuttle's operation, however, was that the solid rockets did not always properly seal at these segments, and that hot gases could leak through the rubber O-rings in the seal, called *blow-by*. This eventually led to the explosion of Challenger in 1986. The pre-accident slide picked out by Feynman had declared that while the lack of a secondary seal in a joint (of the solid rocket motor) was "most critical," it was still "safe to continue flying." At the same time, efforts needed to be "accelerated" to eliminate SRM seal erosion (1988, p. 137). During Columbia as well as Challenger, slides were not just used to support technical and operational decisions that led up to the accidents. Even during both post-accident investigations, slides with bulletized presentations were offered as substitutes for technical analysis and data, causing the CAIB (2003), similar to Feynman years before, to grumble that: "The Board views the endemic use of PowerPoint briefing slides instead of technical papers as an illustration of the problematic methods of technical communication at NASA" (p. 191).

The overuse of bullets and slides illustrates the problem of information environments and how studying them can help us understand something about the creation of local rationality in organizational decision making. NASA's bulletization shows how organizational decision makers are configured in an "epistemic niche" (Hoven, 2001). That which decision makers can know is generated by other people, and gets distorted during transmission through a reductionist, abbreviationist medium. (This epistemic niche also has implications for how we can think about culpability, or blameworthiness of decisions and decision makers—see chap. 10.) The narrowness and incompleteness of the niche in which decision makers find themselves can come across as disquieting to retrospective observers, including people inside and outside the organization. It was after the Columbia accident that the Mission Management Team "admitted that the analysis used to continue flying was, in a word, 'lousy.' This admission—that the rationale to fly was rubber-stamped—is, to say the least, unsettling" (CAIB, 2003, p. 190). "Unsettling" it may be, and probably is—in hindsight. But from the inside, people in organizations do not spend a professional life making "unsettling" decisions. Rather, they do mostly normal work. Again, how can a manager see a "lousy" process to evaluate flight safety as normal, as not something that is worthy reporting or repairing? How could this process be normal? The CAIB (2003) itself found clues to answers in pressures of scarcity and competition:

> The Flight Readiness process is supposed to be shielded from outside influence, and is viewed as both rigorous and systematic. Yet the Shuttle Program is inevitably influenced by external factors, including, in the case of STS-107, schedule demands. Collectively, such factors shape how the Program establishes mission schedules and sets budget priorities, which affects safety oversight, workforce levels, facility maintenance, and contractor workloads. Ultimately, external expectations and pressures impact even data collection, trend analysis, information development, and the reporting and disposition of anomalies. These realities contradict NASA's optimistic belief that preflight reviews provide true safeguards against unacceptable hazards. (2003, p. 191)

Perhaps there is no such thing as "rigorous and systematic" decision making based on technical expertise alone. Expectations and pressures, budget priorities and mission schedules, contractor workloads, and workforce levels all impact technical decision making. All these factors determine and constrain what will be seen as possible and rational courses of action at the time. This dresses up the epistemic niche in which decision makers find themselves in hues and patterns quite a bit more varied than dry technical data alone. But suppose that some decision makers would see through all these dressings on the inside of their epistemic niche, and alert others to it. Tales of such whistleblowers exist. Even if the imperfection of an epistemic niche (the information environment) would be seen and acknowledged from the inside at the time, that still does not mean that it warrants change or improvement. The niche, and the way in which people are configured in it, answers to other concerns and pressures that are active in the organization—efficiency and speed of briefings and decision-making processes, for example. The impact of this imperfect information, even if acknowledged, is underestimated because seeing the side effects, or the connections to real risk, quickly glides outside the computational capability of organizational decision makers and mechanisms at the time.

Studying information environments, how they are created, sustained, and rationalized, and in turn how they help support and rationalize complex and risky decisions, is one route to understanding organizational sensemaking. More will be said on these processes of sensemaking elsewhere in this book. It is a way of making what sociologists call the *macro–micro connection*. How is it that global pressures of production and scarcity find their way into local decision niches, and how is it that they there exercise their often invisible but powerful influence on what people think and prefer; what people then and there see as rational or unremarkable? Although the intention was that NASA's flight safety evaluations be shielded from those external pressures, these pressures nonetheless seeped into even the collection of data, analysis of trends and reporting of anomalies. The information environments thus created for decision makers were continuously and insidiously tainted by pressures of production and scarcity (and in which organization are they not?), prerationally influencing the way people saw the world. Yet even this "lousy" process was considered "normal"—normal or inevitable enough, in any case, to not warrant the expense of energy and political capital on trying to change it. Drift into failure can be the result.

ENGINEERING RESILIENCE INTO ORGANIZATIONS

All open systems are continually adrift inside their safety envelopes. Pressures of scarcity and competition, the intransparency and size of complex systems, the patterns of information that surround decision makers, and the incrementalist nature of their decisions over time, can cause systems to drift into failure. Drift is generated by normal processes of reconciling differential pressures on an organization (efficiency, capacity utilization, safety) against a background of uncertain technology and imperfect knowledge. Drift is about incrementalism contributing to extraordinary events, about the transformation of pressures of scarcity and competition into organizational mandates, and about the normalization of signals of danger so that organizational goals and supposedly normal assessments and decisions become aligned. In safe systems, the very processes that normally guarantee safety and generate organizational success, can also be responsible for organizational demise. The same complex, intertwined sociotechnical life that surrounds the operation of successful technology, is to a large extent responsible for its potential failure. Because these processes are normal, because they are part and parcel of normal, functional organizational life, they are difficult to identify and disentangle. The role of these invisible and unacknowledged forces can be frightening. Harmful consequences can occur in organizations constructed to prevent them. Harmful consequences can occur even when everybody follows the rules (Vaughan, 1996).

The direction in which drift pushes the operation of the technology can be hard to detect, also or perhaps especially for those on the inside. It can be even harder to stop. Given the diversity of forces (political, financial, and economic pressures, technical uncertainty, incomplete knowledge, fragmented problem-solving processes) both on the inside and outside, the large, complex sociotechnical systems that operate some of our most hazardous technologies today seem capable of generating an obscure energy and drift of their own, relatively impervious to outside inspection or inside control.

Recall that, in normal flight, the jackscrew assembly of an MD-80 is supposed to carry a load of about 5,000 pounds. But in effect this load was borne by a leaky, porous, continuously changing system of ill-taught and impractical procedures delegated to operator level that routinely, but forever unsuccessfully, tried to close the gap between production and operation, between making and maintaining. Five thousand pounds of load on a loose and varying collection of procedures and practices were slowly, incrementally grinding their way through the nut threads. It was the sociotechnical system designed to support and protect the uncertain technology, not the mechanical part, that had to carry the load. It gave. The accident report acknowledged that eliminating the risk of single catastrophic failures may not always be possible through design (as design is a reconciliation between irreconcilable constraints). It concluded that "when practicable design alternatives do not exist, a comprehensive systemic maintenance and inspection process is necessary" (NTSB, 2002, p. 180). The conclusion, in other words, was to have a nonredundant system (the single jackscrew and torque tube) made redundant through an organizational-regulatory conglomerate of maintenance and airworthiness checking. The report was forced to conclude that the last resort should be a countermeasure that it had just spent 250 pages proving does not work.

Drifting into failure poses a substantial risk for safe systems. Recognizing and redirecting drift is a challenge that lies before any organization at the 10^{-7} safety frontier. No transportation system in use today has yet broken through this barrier, and success in breaching the asymptote in progress on safety is not likely to come with extensions of current, mechanistic, structuralist approaches. Safety is an emergent property, and its erosion is not about the breakage or lack of quality of single components. This makes the conflation of quality and safety management counterproductive. Many organizations have quality and safety management wrapped up in one function or department. Yet managing quality is about single components, about seeing how they meet particular specifications, about removing or repairing defective components. Managing safety has little to do anymore with single components. An entirely different level of understanding, an entirely different vocabulary is needed to understand safety, in contrast to quality.

Drifting into failure is not so much about breakdowns or malfunctioning of components, as it is about an organization not adapting effectively to cope with the complexity of its own structure and environment. Organizational resilience is not a property, it is a capability: A capability to recognize the boundaries of safe operations, a capability to steer back from them in a controlled manner, a capability to recover from a loss of control if it does occur. This means that human factors and system safety must find new ways of engineering resilience into organizations, of equipping organizations with a capability to recognize, and recover from, a loss of control. How can an organization monitor its own adaptations (and how these bound the rationality of decision makers) to pressures of scarcity and competition, while dealing with imperfect knowledge and uncertain technology? How can an organization become aware, and remain aware, of its models of risk and danger? Organizational resilience is about finding means to invest in safety even under pressures of scarcity and competition, because that may be when such investments are needed most. Preventing a drift into failure requires a different kind of organizational monitoring and learning. It means fixing on higher order variables, adding a new level of intelligence and analysis to the incident reporting and error counting that is done today. More will be said about this in the following chapters.

Chapter 3

Why Are Doctors More Dangerous Than Gun Owners?

There are about 700,000 physicians in the United States. The U.S. Institute of Medicine estimates that each year between 44,000 and 98,000 people die as a result of medical errors (Kohn, Corrigan, & Donaldson, 1999). This makes for a yearly accidental death rate per doctor of between 0.063 and 0.14. In other words, up to one in seven doctors will kill a patient each year by mistake. Take gun owners in contrast. There are 80,000,000 gun owners in the United States. Yet their errors lead to "only" 1,500 accidental gun deaths per year. This means that the accidental death rate, caused by gun-owner error, is 0.000019 per gun owner per year. Only about 1 in 53,000 gun owners will kill somebody by mistake. Doctors then, are 7,500 times more likely to kill somebody by mistake. While not everybody has a gun, almost everybody has a doctor (or several doctors), and is thus severely exposed to the human error problem.

As organizations and other stakeholders (e.g., trade and industry groups, regulators) try to assess the "safety health" of their operations, counting and tabulating errors appears to be a meaningful measure. Not only does it provide an immediate, numeric estimate of the probability of accidental death, injury, or any other undesirable event, it also allows the comparison of systems and components in it (this hospital vs. that hospital, this airline vs. that one, this aircraft fleet or pilot vs. that one, these routes vs. those routes). Keeping track of adverse events is thought to provide relatively easy, quick, and accurate access to the internal safety workings of a system. Moreover, adverse events can be seen as the start of—or reason for—deeper probing, in order to search for environmental threats or unfavorable conditions that could be changed to prevent recurrence. There is, of course, also the sheer scientific curiosity of trying to understand different types of adverse events, different types of errors. Categorizing, after all, has been fundamental to science since modern times.

Over the past decades, transportation human factors has endeavored to quantify safety problems and find potential sources of vulnerability and failure. It has spawned a number of error-classification systems. Some classify decision errors together with the conditions that helped produce them; some have a specific goal, for example, to categorize information transfer problems (e.g., instructions, errors during watch change-over briefings, coordination failures); others try to divide error causes into cognitive, social and situational (physical, environmental, ergonomic) factors; yet others attempt to classify error causes along the lines of a linear information-processing or decision-making model, and some apply the Swiss-cheese metaphor (i.e., systems have multiple layers of defense, but all of them have holes) in the identification of errors and vulnerabilities up the causal chain. Error classification systems are used both after an event (e.g., during incident investigations) or for observations of current human performance.

THE MORE WE MEASURE, THE LESS WE KNOW

In pursuing categorization and tabulation of errors, human factors makes a number of assumptions and takes certain philosophical positions. Little of this is made explicit in the description of these methods, yet it carries consequences for the utility and quality of the error count as a measure of safety health and as a tool for directing resources for improvement. Here is an example. In one of the methods, the observer is asked to distinguish between "procedure errors" and "proficiency errors." Proficiency errors are related to a lack of skills, experience, or (recent) practice, whereas procedure errors are those that occur while carrying out prescribed or normative sequences of action (e.g., checklists). This seems straightforward. Yet, as Croft (2001) reported, the following problem confronts the observer: one type of error (a pilot entering a wrong altitude in a flight computer) can legitimately end up in either of two categories of the error counting method (a procedural error or proficiency error). "For example, entering the wrong flight altitude in the flight management system is considered a procedural error. . . . Not knowing how to use certain automated features in an aircraft's flight computer is considered a proficiency error" (p. 77).

If a pilot enters the wrong flight altitude in the flight-management system, is that a procedural or a proficiency issue, or both? How should it be categorized? Thomas Kuhn (1962) encouraged science to turn to creative philosophy when confronted with the inklings of problems in relating theory to observations (as in the problem of categorizing an observation into theoretical classes). It can be an effective way to elucidate and, if necessary, weaken the grip of a tradition on the collective mind, and suggest the basis for a new one. This is certainly appropriate when epistemological questions arise: questions on how we go about knowing what we (think we) know. To understand error classification and some of the problems associated with it, we should try to engage in a brief analysis of the contemporary philosophical tradition that governs human factors research and the worldview in which it takes place.

Realism: Errors Exist: You Can Discover Them With a Good Method

The position human factors takes when it uses observational tools to measure "errors" is a realist one: It presumes that there is a real, objective world with verifiable patterns that can be observed, categorized, and predicted. Errors, in this sense, are a kind of Durkheimian fact. Emile Durkheim, a founding father of sociology, believed that social reality is objectively "out there," available for neutral, impartial empirical scrutiny. Reality exists, truth is worth striving for. Of course, there are obstacles to getting to the truth, and reality can be hard to pin down. Yet pursuing a close mapping or correspondence to that reality is a valid, legitimate goal of theory development. It is that goal, of achieving a close mapping to reality, that governs error-counting methods. If there are difficulties in getting that correspondence, then these difficulties are merely methodological in nature. The difficulties call for refinement of the observational instruments or additional training of the observers.

These presumptions are modernist; inherited from the enlightened ideas of the Scientific Revolution. In the finest of scientific spirits, method is called on to direct the searchlight across empirical reality, more method is called on to correct ambiguities in the observations, and even more method is called on to break open new portions of hitherto unexplored empirical reality, or to bring into focus those portions that so far were vague and elusive. Other labels that fit such an approach to empirical reality could include *positivism*, which holds that the only type of knowledge worth bothering with is that which is based directly on experience. Positivism is associated with the doctrine of Auguste Comte: The highest, purest (and perhaps only true) form of knowledge is a simple description of sensory phenomena. In other words, if an observer sees an error, then there was an error. For example, the pilot failed to arm the spoilers. This error can then be written up and categorized as such.

But *positivist* has obtained a negative connotation, really meaning "bad" when it comes to social science research. Instead, a neutral way of describing the position of error-counting methods is realist, if naively so. Operating from a realist stance, researchers are concerned with validity (a measure of that correspondence they seek) and reliability. If there is a reality that can be captured and described objectively by outside observers, then it is also possible to generate converging evidence with multiple observers, and consequently achieve agreement about the nature of that reality. This means reliability: Reliable contact has been made with empirical reality, generating equal access and returns across observations and observers. Error counting methods rely on this too: It is possible to tabulate errors from different observers and different observations (e.g., different flights or airlines) and build a common database that can be used as some kind of aggregate norm against which new and existing entrants can be measured.

But absolute objectivity is impossible to obtain. The world is too messy for that, phenomena that occur in the empirical world too confounded, and methods forever imperfect. It comes as no surprise, then, that error-counting methods have different definitions, and different levels of definitions, for *error*, because error itself is a messy and confounded phenomenon:

• Error as the cause of failure, for example, the pilot's failure to arm the spoilers led to the runway overrun.

• Error as the failure itself: Classifications rely on this definition when categorizing the kinds of observable errors operators can make (e.g., decision errors, perceptual errors, skill-based errors; Shappell & Wiegmann, 2001) and probing for the causes of this failure in processing or performance. According to Helmreich (2000), "Errors result from physiological and psychological limitations of humans. Causes of error include fatigue, workload, and fear, as well as cognitive overload, poor interpersonal communications, imperfect information processing, and flawed decision making" (p. 781).

• Error as a process, or, more specifically, as a departure from some kind of standard: This standard may consist of operating procedures. Violations, whether exceptional or routine (Shappell & Wiegmann), or intentional or unintentional (Helmreich), are one example of error according to the process definition. Depending on what they use as standard, observers of course come to different conclusions about what is an error.

Not differentiating among these different possible definitions of error is a well-known problem. Is error a cause, or is it a consequence? To the error-counting methods, such causal confounds and messiness are neither really surprising nor really problematic. Truth, after all, can be elusive. What matters is getting the method right. More method may solve problems of method. That is, of course, if these really are problems of method. The modernist would say "yes." "Yes" would be the stock answer from the Scientific Revolution onward. Methodological wrestling with empirical reality, where empirical reality plays hard to catch and proves pretty good at the game, is just that: methodological. Find a better method, and the problems go away. Empirical reality will swim into view, unadulterated.

Did You Really See the Error Happen?

The postmodernist would argue something different. A single, stable reality that can be approached by the best of methods, and described in terms of correspondence with that reality, does not exist. If we describe reality in a particular way (e.g., this was a "procedure error"), then that does not imply any type of mapping onto an objectively attainable external reality—close or remote, good or bad. The postmodernist does not deal in referentials, does not describe phenomena as though they reflect or represent something stable, objective, something "out there." Rather, capturing and describing a phenomenon is the result of a collective generation and agreement of meaning that, in this case, human factors researchers and their industrial counterparts have reached. The reality of a *procedure error*, in other words, is socially constructed. It is shaped by and dependent on models and paradigms of knowledge that have evolved through group consensus. This meaning is enforced and handed down through systems of observer training, labeling and communication of the results, and industry acceptance and promotion. As philosophers like Kuhn (1962) have pointed out, these paradigms of language and thought at some point adopt a kind of self-sustaining energy, or "consensus authority" (Angell & Straub, 1999). If human factors auditors count errors for managers, they, as (putatively scientific) measurers, have to presume that errors exist. But in order to prove that errors exist, auditors have to measure them. In other words, measuring errors becomes the proof of their existence, an existence that was preordained by their measurement. In the end, everyone agrees that counting errors is a good step forward on safety because almost everyone seems to agree that it is a good step forward. The practice is not questioned because few seem to question it. As the postmodernist would argue, the procedural error becomes true (or appears to people as a close correspondence to some objective reality) only because a community of specialists have contributed to the development of the tools that make it appear so, and have agreed on the language that makes it visible. There is nothing inherently true about the error at all. In accepting the utility of error counting, it is likely that industry accepts its theory (and thereby the reality and validity of the observations it generates) on the authority of authors, teachers, and their texts, not because of evidence. In his headline, Croft (2001) announced that researchers have now perfected ways to monitor pilot performance in the cockpit. "Researchers" have "perfected." There is little that an industry can do other than to accept such authority. What alternatives have they, asks Kuhn, or what competence?

Postmodernism sees the "reality" of an observed procedure error as a negotiated settlement among informed participants. Postmodernism has gone beyond *common denominators*. Realism, that product and accompaniment of the Scientific Revolution, assumes that a common denominator can be found for all systems of belief and value, and that we should strive to converge on those common denominators through our (scientific) methods. There is a truth, and it is worth looking for through method. Postmodernism, in contrast, is the condition of coping without such common denominators. According to postmodernism, all beliefs (e.g., the belief that you just saw a procedural error) are constructions, they are not uncontaminated encounters with, or representations of, some objective empirical reality. Postmodernism challenges the entire modernist culture of realism and empiricism, of which error counting methods are but an instance. Postmodernist defiance not only appears in critiques against error-counting but also reverberates throughout universities and especially the sciences (e.g., Capra, 1982). It never comes away unscathed, however. In the words of Varela, Thompson, and Rosch (1991), we suffer from "Cartesian anxiety." We seem to need the idea of a fixed, stable reality that surrounds us, independent of who looks at it. To give up that idea would be to descend into uncertainty, into idealism, into subjectivism. There would be no more groundedness, no longer a set of predetermined norms or standards, only a constantly shifting chaos of individual impressions, leading to relativism and, ultimately, nihilism. Closing the debate on this anxiety is impossible. Even asking which position is more "real" (the modernist or the postmodernist one) is capitulating to (naive) realism. It assumes that there is a reality that can be approximated better either by the modernists or postmodernists.

Was This an Error? It Depends on Who You Ask

Here is one way to make sense of the arguments. Although people live in the same empirical world (actually, the hard-core constructionist would argue that there is no such thing), they may arrive at rather different, yet equally valid, conclusions about what is going on inside of it, and propose different vocabularies and models to capture those phenomena and activities. Philosophers sometimes use the example of a tree. Though at first sight an objective, stable entity in some external reality, separate from us as observers, the tree can mean entirely different things to someone in the logging industry as compared to, say, a wanderer in the Sahara. Both interpretations can be valid because validity is measured in terms of local relevance, situational applicability, and social acceptability—not in terms of correspondence with a real, external world. Among different characterizations of the world there is no *more real* or *more true*. Validity is a function of how the interpretation conforms to the worldview of those to whom the observer makes his appeal. A *procedure error* is a legitimate, acceptable form of capturing an empirical encounter only because there is a consensual system of like-minded coders and consumers who together have agreed on the linguistic label. The appeal falls onto fertile ground.

But the validity of an observation is negotiable. It depends on where the appeal goes, on who does the looking and who does the listening. This is known as *ontological relativism*: There is flexibility and uncertainty in what it means to be in the world or in a particular situation. The ontological relativist submits that the meaning of observing a particular situation depends entirely on what the observer brings to it. The tree is not just a tree. It is a source of shade, sustenance, survival. Following Kant's ideas, social scientists embrace the common experience that the act of observing and perceiving objects (including humans) is not a passive, receiving process, but an active one that engages the observer as much as it changes or affects the observed. This relativism creates the epistemological uncertainty we see in error-counting methods, which, after all, attempt to shoehorn observations into numerical objectivity. Most social observers or error coders will have felt this uncertainty at one time or another. Was this a procedure error, or a proficiency error, or both? Or was it perhaps no error at all? Was this the cause, or was it the consequence? If it is up to Kant, not having felt this uncertainty would serve as an indication of being a particularly obtuse observer. It would certainly not be proof of the epistemological astuteness of either method or error counter. The uncertainty suffered by them is epistemological because it is realized that certainty about what we know, or even about how to know whether we know it or not, seems out of reach. Yet those within the ruling paradigm have their stock answer to this challenge, just as they have it whenever confronted with problems of bringing observations and theories in closer correspondence. More methodological agreement and refinement, including observer training and standardization, may close the uncertainty. Better trained observers will be able to distinguish between a procedure error and proficiency error, and an improvement to the coding categories may also do the job. Similar modernist approaches have had remarkable success for five centuries, so there is no reason to doubt that they may offer routes to some progress even here. Or is there?

Perhaps more method may not solve problems seemingly linked to method. Consider a study reported by Hollnagel and Amalberti (2001), whose purpose was to test an error-measurement instrument. This instrument was designed to help collect data on, and get a better understanding of, air-traffic controller errors, and to identify areas of weakness and find possibilities for improvement. The method asked observers to count errors (primarily error rates per hour) and categorize the types of errors using a taxonomy proposed by the developers. The tool had already been used to pick apart and categorize errors from past incidents, but would now be put to test in a real-time field setting—applied by pairs of psychologists and air-traffic controllers who would study air-traffic control work going on in real time. The observing air traffic controllers and psychologists, both trained in the error taxonomy, were instructed to take note of all the errors they could see.

Despite common indoctrination, there were substantial differences between the numbers of errors each of the two groups of observers noted, and only a very small number of errors were actually observed by both. People watching the same performance, using the same tool to classify behavior, came up with totally different error counts. Closer inspection of the score sheets revealed that the air-traffic controllers and psychologists tended to use different subsets of the error types available in the tool, indicating just how negotiable the notion of error is: The same fragment of performance means entirely different things to two different (but similarly trained and standardized) groups of observers. Air-traffic controllers relied on external working conditions (e.g., interfaces, personnel and time resources) to refer to and categorize errors, whereas psychologists preferred to locate the error somewhere in presumed quarters of the mind (e.g., working memory) or in some mental state (e.g., attentional lapses). Moreover, air-traffic controllers who actually did the work could tell both groups of error coders that they both had it wrong. Debriefing sessions exposed how many observed errors were not errors at all to those said to have committed them, but rather normal work, expressions of deliberate strategies intended to manage problems or foreseen situations that the error counters had either not seen, or not understood if they had. Croft (2001) reported the same result in observations of cockpit errors: More than half the errors revealed by error counters were never discovered by the flight crews themselves. Some realists may argue that the ability to discover errors that people themselves do not see is a good thing: It confirms the strength or superiority of method. But in Hollnagel and Amalberti's (2001) case, error coders were forced to disavow such claims to epistemological privilege (and embrace ontological relativism instead). They reclassified the errors as normal actions, rendering the score sheets virtually devoid of any error counts. Early transfers of aircraft were not an error, for example, but turned out to correspond to a deliberate strategy connected to a controller's foresight, planning ahead, and workload management. Rather than an expression of weakness, such strategies uncovered sources of robustness that would never have come out, or would even have been misrepresented and mischaracterized, with just the data in the classification tool. Such normalization of actions, which at first appear deviant from the outside, is a critical aspect to really understanding human work and its strengths and weaknesses (see Vaughan, 1996). Without understanding such processes of normalization, it is impossible to penetrate the situated meaning of errors or violations.

Classification of errors crumbles on the inherent weakness of the naive realism that underlies it. The realist idea is that errors are "out there," that they exist and can be observed, captured, and documented independently of the observer. This would mean that it makes no difference who does the observing (which it patently does). Such presumed realism is naive because all observations are ideational—influenced (or made possible in the first place) to a greater or lesser extent by who is doing the observing and by the worldview governing those observations. Realism does not work because it is impossible to separate the observer from the observed. Acknowledging some of these problems, the International Civil Aviation Organization (ICAO, 1998) has called for the development of human performance data-collection methods that do not rely on subjective assessments. But is this possible? Is there such a thing as an objective observation of another human's behavior?

The Presumed Reality of Error

The test of the air-traffic control error counting method reveals how "an action should not be classified as an 'error' only based on how it appears to an observer" (Hollnagel & Amalberti, 2001, p. 13). The test confirms ontological relativism.

Yet sometimes the observed "error" should be entirely non-controversial, should it not? Take the spoiler example from chapter 1. The flight crew forgot to arm the spoilers. They made a mistake. It was an error. You can apply the new view to human error, and explain all about context and situation and mitigating factors. Explain why they did not arm the spoilers, but that they did not arm the spoilers is a fact. The error occurred. Even multiple different observers would agree on that. The flight crew failed to arm the spoilers. How can one not acknowledge the existence of that error? It is there, it is a fact, staring us in the face.

But what is a fact? Facts always privilege the ruling paradigm. Facts always favor current interpretations, as they fold into existing constructed renderings of what is going on. Facts actually exist by virtue of the current paradigm. They can neither be discovered nor given meaning without it. There is no such thing as observations without a paradigm; research in the absence of a particular worldview is impossible. In the words of Paul Feyerabend (1993, p. 11): "On closer analysis, we even find that science knows no 'bare facts' at all, but that the 'facts' that enter our knowledge are already viewed in a certain way and are, therefore, essentially ideational." Feyerabend called the idea that facts are available independently and can thereby objectively favor one theory over another, the *autonomy principle* (p. 26). The autonomy principle asserts that the facts that are available as empirical content of one theory (e.g., procedural errors as facts that fit the threat and error model) are objectively available to alternative theories too. But this does not work. As the spoiler example from chapter 1 showed, errors occur against and because of a background, in this case a background so systemic, so structural, that the original human error pales against it. The error almost becomes transparent, it is normalized, it becomes invisible. Against this backdrop, this context of procedures, timing, engineering trade-offs, and weakened hydraulic systems, the omission to arm the spoilers dissolves. Figure and ground trade places: No longer is it the error that is really observable or even at all interesting. With deeper investigation, ground becomes figure. The backdrop begins to take precedence as the actual story, subsuming, swallowing the original error. No longer can the error be distinguished as a singular, failed decision moment. Somebody who applies a theory of naturalistic decision making will not see a procedure error. What will be seen instead is a continuous flow of actions and assessments, coupled and mutually cued, a flow with nonlinear feedback loops and interactions, inextricably embedded in a multilayered evolving context. Human interaction with a system, in other words, is seen as a continuous control task. Such a characterization is hostile to the digitization necessary to fish out individual human errors.

Whether individual errors can be seen depends on the theory used. There are no objective observations of facts. Observers in error counting are themselves participants, participating in the very creation of the observed fact, and not just because they are there, looking at how other people are working. Of course, through their sheer presence, error counters probably distort people's normal practice, perhaps turning situated performance into a mere window-dressed posture. More fundamentally, however, observers in error counting are participants, because the facts they see would not exist without them. They are created through the method. Observers are participants because it is impossible to separate observer and object.

None of this, by the way, makes the procedure error less real to those who observe it. This is the whole point of ontological relativism. But it does mean that the autonomy principle is false. Facts are not stable aspects of an independent reality, revealed to scientists who wield the right instruments and methods. The discovery and description of every fact is dependent on a particular theory. In the words of Einstein, it is the theory that determines what can be seen. Facts are not available "out there," independent of theory. To suppose that a better theory should come along to account for procedure errors in a way that more closely matches reality is to stick with a model of scientific progress that was disproved long ago. It follows the idea that theories should not be dismissed until there are compelling reasons to do so, and compelling reasons arise only because there is an overwhelming number of facts that disagree with the theory. Scientific work, in this idea, is the clean confrontation of observed fact with theory. But this is not how it works, for those facts do not exist without the theory.

Resisting Change: The Theory Is Right. Or Is It?

The idea of scientific (read: theoretical) progress through the accumulation of observed disagreeing facts that ultimately manage to topple a theory also does not work because counterinstances (i.e., facts that disagree with the theory) are not seen as such. Instead, if observations reveal counterinstances (such as errors that resist unique classification in any of the categories of the error-counting method), then researchers tend to see these as further puzzles in the match between observation and theory (Kuhn, 1962)—puzzles that can be addressed by further refinement of their method.

Counterinstances, in other words, are not seen as speaking against the theory. According to Kuhn (1962), one of the defining responses to paradigmatic crisis is that scientists do not treat anomalies as counterinstances, even though that is what they are. It is extremely difficult for people to renounce the paradigm that has led them into a crisis. Instead, the epistemological difficulties suffered by error-counting methods (Was this a cause or a consequence? Was this a procedural or a proficiency error?) are dismissed as minor irritants and reasons to engage in yet more methodological refinement consonant with the current paradigm.

Neither scientists nor their supporting communities in industry are willing to forego a paradigm until and unless there is a viable alternative ready to take its place. This is among the most sustained arguments surrounding the continuation of error counting: Researchers engaging in error classification are willing to acknowledge that what they do is not perfect, but vow to keep going until shown something better. And industry concurs. As Kuhn pointed out, the decision to reject one paradigm necessarily coincides with the acceptance of another.

Proposing a viable alternative theory that can deal with its own facts, however, is exceedingly difficult, and has proven to be so even historically (Feyerabend, 1993). Facts, after all, privilege the status quo. Galileo's telescopic observations of the sky generated observations that motivated an alternative explanation about the place of the earth in the universe. His observations favored the Copernican heliocentric interpretation (where the earth goes around the sun) over the Ptolomeic geocentric one (where the sun goes around the earth). The Copernican interpretation, however, was a worldview away from what was currently accepted, and many doubted Galileo's data as a valid empirical window on that heliocentric reality. People were highly suspicious of the new instrument: Some asked Galileo to open up his telescope to prove that there was no little moon hiding inside of it. How, otherwise, could the moon or any other celestial body be seen so closely if it was not itself hiding in the telescope? One problem was that Galileo did not offer a theoretical explanation for why this could be so, and why the telescope was supposed to offer a better picture of the sky than the naked eye. He could not, because relevant theories (optica) were not yet well developed. Generating better data (like Galileo did), and developing entirely new methods for better access to these data (such as a telescope), does in itself little to dislodge an established theory that allows people to see the phenomenon with their naked eye and explain it with their common sense. Similarly, people see the error happen with their naked eye, even without the help of an error-classification method: The pilot fails to arm the spoilers. Even their common sense confirms that this is an error. The sun goes around the earth. The earth is fixed. The Church was right, and Galileo was wrong. None of his observed facts could prove him right, because there was no coherent set of theories ready to accommodate his facts and give them meaning. The Church was right, as it had all the facts. And it had the theory to deal with them.

Interestingly, the Church kept closer to reason as it was defined at the time. It considered the social, political, and ethical implications of Galileo's alternatives and deemed them too risky to accept—certainly on the grounds of tentative, rickety evidence. Disavowing the geocentric idea would be disavowing Creation itself, removing the common ontological denominator of the past millennium and severely undermining the authority and political power the Church derived from it. Error-classification methods too, guard a piece of rationality that most people in industry and elsewhere would be loathe to see disintegrate. Errors occur, they can be distinguished objectively. Errors can be an indication of unsafe performance. There is good performance and bad performance; there are identifiable causes for why people perform well or less well and for why failures happen. Without such a supposedly factual basis, without such hopes of an objective rationality, traditional and well-established ways for dealing with threats to safety and trying to create progress could collapse. Cartesian anxiety would grip the industry and research community. How can we hold people accountable for mistakes if there are no "errors"? How can we report safety occurrences and maintain expensive incident-reporting schemes if there are no errors? What can we fix if there are no causes for adverse events? Such questions fit a broader class of appeals against relativism. Postmodernism and relativism, according to their detractors, can lead only to moral ambiguity, nihilism and lack of structural progress. We should instead hold onto the realist status quo, and we can, for most observed facts still seem to privilege it. Errors exist. They have to.

To the naive realist, the argument that errors exist is not only natural and necessary, it is also quite impeccable, quite forceful. The idea that errors do not exist, in contrast, is unnatural, even absurd. Those within the established paradigm will challenge the sheer legitimacy of questions raised about the existence of errors, and by implication even the legitimacy of those who raise the questions: "Indeed, there are some psychologists who would deny the existence of errors altogether. We will not pursue that doubtful line of argument here" (Reason & Hobbs, 2003, p. 39).

Because the current paradigm judges it absurd and unnatural, the question about whether errors exist is not worth pursuing: It is doubtful and unscientific—and in the strictest sense (when scientific pursuits are measured and defined within the ruling paradigm), that is precisely what it is. If some scientists do not succeed in bringing statement and fact into closer agreement (they do not see a procedure error where others would), then this discredits the scientist rather than the theory. Galileo suffered from this too. It was the scientist who was discredited (for a while at least), not the prevailing paradigm. So what does he do? How does Galileo proceed once he introduces an interpretation so unnatural, so absurd, so countercultural, so revolutionary? What does he do when he notices that even the facts are not (interpreted to be) on his side? As Feyerabend (1993) masterfully described it, Galileo engaged in propaganda and psychological trickery. Through imaginary conversations between Sagredo, Salviati, and Simplicio, written in his native tongue rather than in Latin, he put the ontological uncertainty and epistemological difficulty of the geocentric interpretation on full display. The sheer logic of the geocentric interpretation fell apart whereas that of the heliocentric interpretation triumphed. Where the appeal to empirical facts failed (because those facts will still be forced to fit the prevailing paradigm rather than its alternative), an appeal to logic may still succeed. The same is true for error counting and classification. Just imagine this conversation:

Simplicio: Errors result from physiological and psychological limitations of humans. Causes of error include fatigue, workload, and fear, as well as cognitive overload, poor interpersonal communications, imperfect information processing, and flawed decision making.

Sagredo: But are errors in this case not simply the result of other errors? Flawed decision making would be an error. But in your logic, it causes an error. What is the error then? And how can we categorize it?

Simplicio: Well, but errors are caused by poor decisions, failures to adhere to brief, failures to prioritize attention, improper procedure, and so forth.

Sagredo: This appears to be not causal explanation, but simply relabeling. Whether you say error, or poor decision, or failure to prioritize attention, it all still sounds like error, at least when interpreted in your worldview. And how can one be the cause of the other to the exclusion of the other way around? Can errors cause poor decisions just like poor decisions cause errors? There is nothing in your logic that rules this out, but then we end up with a tautology, not an explanation.

And yet, such arguments may not help either. The appeal to logic may fail in the face of overwhelming support for a ruling paradigm—support that derives from consensus authority, from political, social, and organizational imperatives rather than a logical or empirical basis (which is, after all, pretty porous). Even Einstein expressed amazement at the common reflex to rely on measurements (e.g., error counts) rather than logic and argument: " 'Is it not really strange,' he asked in a letter to Max Born, 'that human beings are normally deaf to the strongest of argument while they are always inclined to overestimate measuring accuracies?' " (Feyerabend, 1993, p. 239). Numbers are strong. Arguments are weak. Error counting is good because it generates numbers, it relies on accurate measurements (recall Croft, 2001, who announced that "researchers" have "perfected" ways to monitor pilot performance), rather than on argument.

In the end, no argument, none of this propaganda or psychological trickery can serve as a substitute for the development of alternative theory, nor did it in Galileo's case. The postmodernists are right and the realists are wrong: Without a paradigm, without a worldview, there are no facts. People will reject no theory on the basis of argument or logic alone. They need another to take its place. A paradigmatic interregnum would produce paralysis. Suspended in a theoretical vacuum, researchers would no longer be able to see facts or do anything meaningful with them.

So, considering the evidence, what should the alternative theory look like? It needs to come with a superior explanation of performance variations, with an interpretation that is sensitive to the situatedness of the performance it attempts to capture. Such a theory sees no errors, but rather performance variations—inherently neutral changes and adjustments in how people deal with complex, dynamic situations. This theory will resist coming in from the outside, it will avoid judging other people from a position external to how the situation looked to the subject inside of it. The outlines of such a theory are developed further in various places in this book.

SAFETY AS MORE THAN ABSENCE OF NEGATIVES

First, though, another question: Why do people bother with error counts in the first place? What goals do they hope these empirical measures help them accomplish, and are there better ways to achieve those goals? A final aim of error counting is to help make progress on safety, but this puts the link between errors and safety on trial. Can the counting of negatives (e.g., these errors) say anything useful about safety? What does the quantity measured (errors) have to do with the quality managed (safety)? Error-classification methods assume a close mapping between these two, and assume that an absence or reduction of errors is synonymous with progress on safety. By treating safety as positivistically measurable, error counting may be breathing the scientific spirit of a bygone era. Human performance in the laboratory was once gauged by counting errors, and this is still done when researchers test limited, contrived task behavior in spartan settings. But how well does this export to natural settings where people carry out actual complex, dynamic, and interactive work, where determinants of good and bad outcomes are deeply confounded?

It may not matter. The idea of a realist count is compelling to industry for the same reasons that any numerical performance measurement is. Managers get easily infatuated with "balanced scorecards" or other faddish figures of performance. Entire business models depend on quantifying performance results, so why not quantify safety? Error counting becomes yet another quantitative basis for managerial interventions. Pieces of data from the operation that have been excised and formalized away from their origin can be converted into graphs and bar charts that subsequently form the inspiration for interventions. This allows managers, and their airlines, to elaborate their idea of control over operational practice and its outcomes. Managerial control, however, exists only in the sense of purposefully formulating and trying to influence the intentions and actions of operational people (Angell & Straub, 1999). It is not the same as being in control of the consequences (by which safety ultimately gets measured industry-wide), because for that the real world is too complex and operational environments too stochastic (e.g., Snook, 2000).

There is another tricky aspect of trying to create progress on safety through error counting and classification. This has to do with not taking context into regard when counting errors. Errors, according to realist interpretations, represent a kind of equivalent category of bad performance (e.g., a failure to meet one's objective or intention), no matter who commits the error or in what situation. Such an assumption has to exist, otherwise tabulation becomes untenable. One cannot (or should not) add apples and oranges, after all. If both apples and oranges are entered into the method (and, given that the autonomy principle is false, error-counting methods do add apples and oranges), silly statistical tabulations that claim doctors are 7,500 times more dangerous than gun owners can roll out the other end. As Hollnagel and Amalberti (2001) showed, attempts to map situated human capabilities such as decision making, proficiency, or deliberation onto discrete categories are doomed to be misleading. They cannot cope with the complexity of actual practice without serious degeneration (Angell & Straub, 1999). Error classification disembodies data. It removes the context that helped produce the behavior in its particular manifestation. Such disembodiment may actually retard understanding. The local rationality principle (people's behavior is rational when viewed from the inside of their situations) is impossible to maintain when context is removed from the controversial action. And error categorization does just that: It removes context. Once the observation of some kind of error is tidily locked away into some category, it has been objectified, formalized away from the situation that brought it forth. Without context, there is no way to reestablish local rationality. And without local rationality, there is no way to understand human error. And without understanding human error, there may be no way to learn how to create progress on safety.

Safety as Reflexive Project

Safety is likely to be more than the measurement and management of negatives (errors), if it is that at all. Just as errors are epistemologically elusive (How do you know what you know? Did you really see a procedure error? Or was it a proficiency error?), and ontologically relativist (what it means "to be" and to perform well or badly inside a particular situation is different from person to person), the notion of safety may similarly lack an objective, common denominator. The idea behind measuring safety through error counts is that safety is some kind of objective, stable (and perhaps ideal) reality, a reality that can be measured and reflected, or represented, through method. But does this idea hold? Rochlin (1999, p. 1550), for example, proposed that safety is a "constructed human concept" and others in human factors have begun to probe how individual practitioners construct safety, by assessing what they understand risk to be, and how they perceive their ability of managing challenging situations (e.g., Orasanu, 2001). A substantial part of practitioners' construction of safety turns out to be reflexive, assessing the person's own competence or skill in maintaining safety across different situations. Interestingly, there may be a mismatch between risk salience (how critical a particular threat to safety was perceived to be by the practitioner) and frequency of encounters (how often these threats to safety are in fact met in practice). The safety threats deemed most salient were the ones least frequently dealt with (Orasanu, 2001). Safety is more akin to a reflexive project, sustained through a revisable narrative of self-identity that develops in the face of frequently and less frequently en-countered risks. It is not something referential, not something that is objec-tively "out there" as a common denominator, open to any type of approxi-mation by those with the best methods. Rather, safety may be reflexive: something that people relate to themselves.

The numbers produced by error counts are a logical endpoint of a struc-tural analysis that focuses on (supposed) causes and consequences, an anal-ysis that defines risk and safety instrumentally, in terms of minimizing er-rors and presumably measurable consequences. A second, more recent approach is more socially and politically oriented, and places emphasis on representation, perception, and interpretation rather than on structural features (Rochlin, 1999). The managerially appealing numbers generated by error counts do not carry any of this reflexivity, none of the nuances of what it is to "be there," doing the work, creating safety on the line. What it is to be there ultimately determines safety (as outcome): People's local ac-tions and assessments are shaped by their own perspectives. These in turn are embedded in histories, rituals, interactions, beliefs and myths, both of people's organization and organizational subculture and of them as indi-viduals. This would explain why good, objective, empirical indicators of so-cial and organizational definitions of safety are difficult to obtain. Opera-tors of reliable systems "were expressing their evaluation of a positive state mediated by human action, and that evaluation reflexively became part of the state of safety they were describing" (Rochlin, 1999, p. 1550). In other words, the description itself of what safety means to an individual operator is a part of that very safety, dynamic and subjective. "Safety is in some sense a story a group or organization tells about itself and its relation to its task environment" (Rochlin, p. 1555).

Can We Measure Safety?

But how does an organization capture what groups tell about themselves; how does it pin down these stories? How can management measure a medi-ated, reflexive idea? If not through error counts, what can an organization look for in order to get some measure of how safe it is? Large recent acci-dents provide some clues of where to start looking (e.g., Woods, 2003). A main source of residual risk in otherwise safe transportation systems is the drift into failure described in chapter 2. Pressures of scarcity and competi-tion narrow an organization's focus on goals associated with production. With an accumulating base of empirical success (i.e., no accidents, even if safety is increasingly traded off against other goals such as maximizing profit or capacity utilization), the organization, through its members' mul-tiple little and larger daily decisions, will begin to believe that past success is a guarantee of future safety, that historical success is a reason for confidence that the same behavior will lead to the same (successful) outcome the next time around. The absence of failure, in other words, is taken as evidence that hazards are not present, that countermeasures already in place are effective. Such a model of risk is embedded deeply in the reflexive stories of safety that Rochlin (1999) talked about, and it can be made explicit only through qualitative investigations that probe the interpretative aspect of situated human assessments and actions. Error counts do little to elucidate any of this. More qualitative studies could reveal how currently traded models of risk may increasingly be at odds with the actual nature and proximity of hazard, though it may of course be difficult to establish the objective, or ontologically absolutist, presence of hazard.

Particular aspects of how organization members tell or evaluate safety stories, however, can serve as markers. Woods (2003, p. 5), for example, has called one of these markers "distancing through differencing." In this process, organizational members look at other failures and other organizations as not relevant to them and their situation. They discard other events because they appear at the surface to be dissimilar or distant. Discovering this through qualitative inquiry can help specify how people and organizations reflexively create their idea, their story of safety. Just because the organization or section has different technical problems, different managers, different histories, or can claim to already have addressed a particular safety concern revealed by the event, does not mean that they are immune to the problem. Seemingly divergent events can represent similar underlying patterns in the drift towards hazard. High-reliability organizations characterize themselves through their preoccupation with failure: continually asking themselves how things can go wrong and could have gone wrong, rather than congratulating themselves on the fact that things went right. Distancing through differencing means underplaying this preoccupation. It is one way to prevent learning from events elsewhere, one way to throw up obstacles in the flow of safety-related information.

Additional processes that can be discovered include to what extent an organization resists oversimplifying interpretations of operational data, whether it defers to expertise and expert judgment rather than managerial imperatives. Also, it could be interesting to probe to what extent problem-solving processes are fragmented across organizational departments, sections, or subcontractors. The 1996 ValuJet accident, where flammable oxygen generators were placed in an aircraft cargo hold without shipping caps, subsequently burning down the aircraft, was related to a web of subcontractors that together made up the virtual airline of ValuJet. Hundreds of people within even one subcontractor logged work against the particular ValuJet aircraft, and this subcontractor was only one of many players in a network of organizations and companies tasked with different aspects of running (even constituting) the airline. Relevant maintenance parts (among them the shipping caps) were not available at the subcontractor, ideas of what to do with expired oxygen canisters were generated ad hoc in the absence of central guidance, and local understandings for why shipping caps may have been necessary were foggy at best. With work and responsibility for it distributed among so many participants, nobody may have been able anymore to see the big picture, including the regulator. Nobody may have been able to recognize the gradual erosion of safety constraints on the design and operation of the original system.

If safety is a reflexive project rather than an objective datum, human factors researchers must develop entirely new probes for measuring the safety health of an organization. Error counts do not suffice. They uphold an illusion of rationality and control, but may offer neither real insight nor productive routes for progress on safety. It is, of course, a matter of debate whether the vaguely defined organizational processes that could be part of new safety probes (e.g., distancing through differencing, deference to expertise, fragmentation of problem-solving, incremental judgments into disaster) are any more real than the errors from the counting methods they seek to replace or augment. But then, the reality of these phenomena is in the eye of the beholder: Observer and observed cannot be separated; object and subject are largely indistinguishable. The processes and phenomena are real enough to those who look for them and who wield the theories to accommodate the results. Criteria for success may lie elsewhere, for example in how well the measure maps onto past evidence of precursors to failure. Yet even such mappings are subject to paradigmatic interpretations of the evidence base. Indeed, consonant with the ontological relativity of the age human factors has now entered, the debate can probably never be closed. Are doctors more dangerous than gun owners? Do errors exist? It depends on who you ask.

The real issue, therefore, lies a step away from the fray, a level up, if you will. Whether we count errors as Durkheimian fact on the one hand or see safety as a reflexive project on the other, competing premises and practices reflect particular models of risk. These models of risk are interesting not because of their differential abilities to access empirical truth (because that may all be relative), but because of what they say about us, about human factors and system safety. It is not the monitoring of safety that we should simply pursue, but the monitoring of that monitoring. If we want to make progress on safety, one important step is to engage in such metamonitoring, to become better aware of the models of risk embodied in our assumptions and approaches to safety.

Chapter 4

Don't Errors Exist?

Human factors as a discipline takes a very realist view. It lives in a world of real things, of facts and concrete observations. It presumes the existence of an external world in which phenomena occur that can be captured and described objectively. In this world there are errors and violations, and these errors and violations are quite real. The flight-deck observer from chapter 3, for example, would see that pilots do not arm the spoilers before landing and marks this up as an error or a procedural violation. The observer considers his observation quite true, and the error quite real.

Upon discovering that the spoilers had not been armed, the pilots themselves too may see their omission as an error, as something that they missed but should not have missed. But just as it did for the flight-deck observer, the error becomes real only because it is visible from outside the stream of experience. From the inside of this stream, while things are going on and work is being accomplished, there is no error. In this case there are only procedures that get inadvertently mangled through the timing and sequence of various tasks. And not even this gets noticed by those applying the procedures.

Recall how Feyerabend (1993) pointed out that all observations are ideational, that facts do not exist without an observer wielding a particular theory that tells him or her what to look for. Observers are not passive recipients, but active creators of the empirical reality they encounter. There is no clear separation between observer and observed. As said in chapter 3, none of this makes the error any less real to those who observe it. But it does not mean that the error exists out there, in some independent empirical universe. This was the whole point of ontological relativism: What it means to be in a particular situation and make certain observations is quite flexible and connected systematically to the observer. None of the possible world-views can be judged superior or privileged uniquely by empirical data about the world, because objective, impartial access to that world is impossible. Yet in the pragmatic and optimistically realist spirit of human factors, error counting methods have gained popularity by selling the belief that such impartial access is possible. The claim to privileged access lies (as modernism and Newtonian science would dictate) in method. The method is strong enough to discover errors that the pilots themselves had not seen.

Errors appear so real when we step or set ourselves outside the stream of experience in which they occur. They appear so real to an observer sitting behind the pilots. They appear so real to even the pilot himself after the fact. But why? It cannot be because the errors are real, since the autonomy principle has been proven false. As an observed fact, the error only exists by virtue of the observer and his or her position on the outside of the stream of experience. The error does not exist because of some objective empirical reality in which it putatively takes place, since there is no such thing and if there was, we could not know it. Recall the air-traffic control test of chapter 3: Actions, omissions, and postponements related to air-traffic clearances carry entirely different meanings for those on the inside and on the outside of the work experience. Even different observers on the outside cannot agree on a common denominator because they have diverging backgrounds and conceptual looking glasses. The autonomy principle is false: facts do not exist without an observer. So why do errors appear so real?

ERRORS ARE ACTIVE, CORRECTIVE INTERVENTIONS IN HISTORY

Errors are an active, corrective intervention in (immediate) history. It is impossible for us to give a mere chronicle of our experiences: Our assumptions, past experiences and future aspirations cause us to impress a certain organization on that which we just went through or saw. Errors are a powerful way to impose structure onto past events. Errors are a particular way in which we as observers (or even participants) reconstruct the reality we just experienced. Such reconstruction, however, inserts a severe discontinuity between past and present. The present was once an uncertain, perhaps vanishingly improbable, future. Now we see it as the only plausible outcome of a pretty deterministic past. Being able to stand outside an unfolding sequence of events (either as participants from hindsight or as observers from outside the setting) makes it exceedingly difficult to see how unsure we once were (or could have been if we had been in that situation) of what was going to happen. History as seen through the eyes of a retrospective outsider (even if the same observer was a participant in that history not long ago) is substantially different from the world as it appeared to the decision makers of the day. This endows history, even immediate history, with a determinism it lacked when it was still unfolding.

Errors, then, are *ex post facto* constructs. The research base on the hindsight bias contains some of the strongest evidence on this. Errors are not empirical facts. They are the result of outside observers squeezing now-known events into the most plausible or convenient deterministic scheme. In the research base on hindsight, it is not difficult to see how such retrospective restructuring embraces a liberal take on the history it aims to recount. The distance between reality as portrayed by a retrospective observer and as experienced by those who were there (even if these were once the same people) grows substantially with the rhetoric and discourse employed and the investigative practices used. We see a lot of this later in the discussion.

We also look at developments in psychology that have (since not so long ago) tried to get away from the normativist bias in our understanding of human performance and decision making. This intermezzo is necessary because errors and violations do not exist without some norm, even if implied. Hindsight of course has a powerful way of importing criteria or norms from outside people's situated contexts, and highlighting where actual performance at the time fell short. To see errors as *ex post* constructs rather than as objective, observed facts, we have to understand the influence of implicit norms on our judgments of past performance. Doing without errors means doing with normativism. It means that we cannot question the accuracy of insider accounts (something human factors consistently does, e.g., when it asserts a "loss of situation awareness"), as there is no objective, normative reality to hold such accounts up to, and relative to which we can deem them accurate or inaccurate. Reality as experienced by people at the time was reality as it was experienced by them at the time, full stop. It was that experienced world that drove their assessments and decisions, not our (or even their) retrospective, outsider rendering of that experience. We have to use local norms of competent performance to understand why what people did made sense to them at the time.

Finally, an important question we must look ahead to: Why is it that errors fulfill such an important function in our reconstructions of history, of even our own histories? Seeing errors in history may actually have little to do with historical explanation. Rather, it may be about controlling the future. What we see toward the end of this chapter is that the hindsight bias may not at all be about history, and may not even be a bias. Retrospective reconstruction, and the hindsight bias, should not be seen as the primary phenomenon. Rather, it represents and serves a larger purpose, answering a highly pragmatic concern. The almost inevitable urge to highlight past choice moments (where people went the wrong way), the drive to identify errors, is forward looking, not backward looking. The hindsight bias may not be a bias because it is an adaptive response, an oversimplification of history that primes us for complex futures and allows us to project simple models of past lessons onto those futures, lest history repeat itself. This means that retrospective recounting tells us much more about the observer than it does about reality—if there is such an objective thing.

Making Tangled Histories Linear

The hindsight bias (Fischoff, 1975) is one of the most consistent biases in psychology. One effect is that "people who know the outcome of a complex prior history of tangled, indeterminate events, remember that history as being much more determinant, leading 'inevitably' to the outcome they already knew" (Weick, 1995, p. 28). Hindsight allows us to change past indeterminacy and complexity into order, structure, and oversimplified causality (Reason, 1990). As an example, take the turn towards the mountains that a Boeing 757 made just before an accident near Cali, Colombia in 1995. According to the investigation, the crew did not notice the turn, at least not in time (Aeronautica Civil, 1996). What should the crew have seen in order to know about the turn? They had plenty of indications, according to the manufacturer of their aircraft:

> Indications that the airplane was in a left turn would have included the following: the EHSI (Electronic Horizontal Situation Indicator) Map Display (if selected) with a curved path leading away from the intended direction of flight; the EHSI VOR display, with the CDI (Course Deviation Indicator) displaced to the right, indicating the airplane was left of the direct Cali VOR course, the EaDI indicating approximately 16 degrees of bank, and all heading indicators moving to the right. Additionally the crew may have tuned Rozo in the ADF and may have had bearing pointer information to Rozo NDB on the RMDI. (Boeing Commercial Airplane Group, 1996, p. 13)

This is a standard response after mishaps: Point to the data that would have revealed the true nature of the situation. In hindsight, there is an overwhelming array of evidence that did point to the real nature of the situation, and if only people had paid attention to even some of it, the outcome would have been different. Confronted with a litany of indications that could have prevented the accident, we wonder how people at the time could not have known all of this. We wonder how this "epiphany" was missed, why this bloated shopping bag full of revelations was never opened by the people who most needed it.

But knowledge of the critical data comes only with the omniscience of hindsight. We can only know what really was critical or highly relevant once we know the outcome. Yet if data can be shown to have been physically available, we often assume that it should have been picked up by the practitioners in the situation. The problem is that pointing out that something should have been noticed does not explain why it was not noticed, or why it was interpreted differently back then. This confusion has to do with us, not with the people we are investigating. What we, in our reaction to failure, fail to appreciate is that there is a dissociation between data availability and data observability—between what can be shown to have been physically available and what would have been observable given people's multiple interleaving tasks, goals, attentional focus, expectations, and interests. Data, such as the litany of indications in the previous example, do not reveal themselves to practitioners in one big monolithic moment of truth. In situations where people do real work, data can get drip-fed into the operation: a little bit here, a little bit there. Data emerges over time. Data may be uncertain. Data may be ambiguous. People have other things to do too. Sometimes the successive or multiple data bits are contradictory, often they are unremarkable. It is one thing to say how we find some of these data important in hindsight. It is quite another to understand what the data meant, if anything, to the people in question at the time.

The same kind of confusion occurs when we, in hindsight, get an impression that certain assessments and actions point to a common condition. This may be true at first sight. In trying to make sense of past performance, it is always tempting to group individual fragments of human performance that seem to share something, that seem to be connected in some way, and connected to the eventual outcome. For example, "hurry" to land was such a leitmotif extracted from the evidence in the Cali investigation. Haste in turn is enlisted to explain the errors that were made:

> Investigators were able to identify a series of errors that initiated with the flightcrew's acceptance of the controller's offer to land on runway 19 . . . The CVR (Cockpit Voice Recorder) indicates that the decision to accept the offer to land on runway 19 was made jointly by the captain and the first officer in a 4-second exchange that began at 2136:38. The captain asked: "would you like to shoot the one nine straight in?" The first officer responded, "Yeah, we'll have to scramble to get down. We can do it." This interchange followed an earlier discussion in which the captain indicated to the first officer his desire to hurry the arrival into Cali, following the delay on departure from Miami, in an apparent to minimize the effect of the delay on the flight attendants' rest requirements. For example, at 2126:01, he asked the first officer to "keep the speed up in the descent" . . . (This is) evidence of the hurried nature of the tasks performed. (Aeronautica Civil, 1996, p. 29)

In this case the fragments used to build the argument of haste come from over half an hour of extended performance. Outside observers have treated the record as if it were a public quarry to pick stones from, and the accident explanation the building he needs to erect. The problem is that each fragment is meaningless outside the context that produced it: Each fragment has its own story, background, and reasons for being, and when it was produced it may have had nothing to do with the other fragments it is now grouped with. Moreover, behavior takes place in between the fragments. These intermediary episodes contain changes and evolutions in perceptions and assessments that separate the excised fragments not only in time, but also in meaning. Thus, the condition, and the constructed linearity in the story that binds these performance fragments, does not arise from the circumstances that brought each of the fragments forth; it is not a feature of those circumstances. It is an artifact of the outside observer. In the case just described, hurry is a condition identified in hindsight, one that plausibly couples the start of the flight (almost 2 hours behind schedule) with its fatal ending (on a mountainside rather than an airport). Hurry is a retrospectively invoked leitmotif that guides the search for evidence about itself. It leaves the investigator with a story that is admittedly more linear and plausible and less messy and complex than the actual events. Yet it is not a set of findings, but of tautologies.

Counterfactual Reasoning

Tracing the sequence of events back from the outcome—that we as outside observers already know about—we invariably come across joints where people had opportunities to revise their assessment of the situation but failed to do so, where people were given the option to recover from their route to trouble, but did not take it. These are counterfactuals—quite common in accident analysis. For example, "The airplane could have overcome the windshear encounter if the pitch attitude of 15 degrees nose-up had been maintained, the thrust had been set to 1.93 EPR (Engine Pressure Ratio) and the landing gear had been retracted on schedule" (NTSB, 1995, p. 119). Counterfactuals prove what could have happened if certain minute and often utopian conditions had been met.

Counterfactual reasoning may be a fruitful exercise when trying to uncover potential countermeasures against such failures in the future. But saying what people could have done in order to prevent a particular outcome does not explain why they did what they did. This is the problem with counterfactuals. When they are enlisted as explanatory proxy, they help circumvent the hard problem of investigations: finding out why people did what they did. Stressing what was not done (but if it had been done, the accident would not have happened) explains nothing about what actually happened, or why. In addition, counterfactuals are a powerful tributary to the hindsight bias. They help us impose structure and linearity on tangled prior histories. Counterfactuals can convert a mass of indeterminate actions and events, themselves overlapping and interacting, into a linear series of straightforward bifurcations. For example, people could have perfectly executed the go-around maneuver but did not; they could have denied the runway change but did not. As the sequence of events rolls back into time, away from its outcome, the story builds. We notice that people chose the wrong prong at each fork, time and again—ferrying them along inevitably to the outcome that formed the starting point of our investigation (for without it, there would have been no investigation).

But human work in complex, dynamic worlds is seldom about simple dichotomous choices (as in: to err or not to err). Bifurcations are extremely rare—especially those that yield clear previews of the respective outcomes at each end. In reality, choice moments (such as there are) typically reveal multiple possible pathways that stretch out, like cracks in a window, into the ever denser fog of futures not yet known. Their outcomes are indeterminate, hidden in what is still to come. In reality, actions need to be taken under uncertainty and under the pressure of limited time and other resources. What from the retrospective outside may look like a discrete, leisurely two-choice opportunity to not fail, is from the inside really just one fragment caught up in a stream of surrounding actions and assessments.

In fact, from the inside it may not look like a choice at all. These are often choices only in hindsight. To the people caught up in the sequence of events, there was perhaps not any compelling reason to reassess their situation or decide against anything (or else they probably would have) at the point the investigator has now found significant or controversial. They were likely doing what they were doing because they thought they were right, given their understanding of the situation, their pressures. The challenge for an investigator becomes to understand how this may not have been a discrete event to the people whose actions are under investigation. The investigator needs to see how other people's decisions to continue were likely nothing more than continuous behavior—reinforced by their current understanding of the situation, confirmed by the cues they were focusing on, and reaffirmed by their expectations of how things would develop.

Judging Instead of Explaining

When outside observers use counterfactuals, even as explanatory proxy, they themselves often require explanations as well. After all, if an exit from the route to trouble stands out so clearly to outside observers, how was it possible for other people to miss it? If there was an opportunity to recover, to not crash, then failing to grab it demands an explanation. The place where observers often look for clarification is the set of rules, professional standards, and available data that surrounded people's operation at the time, and how people did not see or meet that which they should have seen or met. Recognizing that there is a mismatch between what was done or seen and what should have been done or seen—as per those standards—we easily judge people for not doing what they should have done.

Where fragments of behavior are contrasted with written guidance that can be found to have been applicable in hindsight, actual performance is often found wanting; it does not live up to procedures or regulations. For example, "One of the pilots . . . executed [a computer entry] without having verified that it was the correct selection and without having first obtained approval of the other pilot, contrary to procedures" (Aeronautica Civil, 1996, p. 31). Investigations invest considerably in organizational archeology so that they can construct the regulatory or procedural framework within which the operations took place, or should have taken place. Inconsistencies between existing procedures or regulations and actual behavior are easy to expose when organizational records are excavated after the fact and rules uncovered that would have fit this or that particular situation.

This is not, however, very informative. There is virtually always a mismatch between actual behavior and written guidance that can be located in hindsight. Pointing out a mismatch sheds little light on the why of the behavior in question, and, for that matter, mismatches between procedures and practice are not unique to mishaps. There are also less obvious or undocumented standards. These are often invoked when a controversial fragment (e.g., a decision to accept a runway change, Aeronautica Civil, 1996, or the decision to go around or not, NTSB, 1995) knows no clear preordained guidance but relies on local, situated judgment. For these cases there are always supposed standards of good practice, based on convention and putatively practiced across an entire industry. One such standard in aviation is "good airmanship," which, if nothing else can, will explain the variance in behavior that had not yet been accounted for.

When micromatching, observers frame people's past assessments and actions inside a world that they have invoked retrospectively. Looking at the frame as overlay on the sequence of events, they see that pieces of behavior stick out in various places and at various angles: a rule not followed here, available data not observed there, professional standards not met over there. But rather than explaining controversial fragments in relation to the circumstances that brought them forth, and in relation to the stream of preceding as well as succeeding behaviors that surrounded them, the frame merely boxes performance fragments inside a world observers now know to be true. The problem is that this after-the-fact-world may have very little relevance to the actual world that produced the behavior under study. The behavior is contrasted against the observer's reality, not the reality surrounding the behavior at the time. Judging people for what they did not do relative to some rule or standard does not explain why they did what they did. Saying that people failed to take this or that pathway—only in hindsight the right one—judges other people from a position of broader insight and outcome knowledge that they themselves did not have. It does not explain a thing yet; it does not shed any light on why people did what they did given their surrounding circumstances. Outside observers have become caught in what William James called the "psychologist's fallacy" a century ago: They have substituted their own reality for the one of their object of study.

The More We Know, the Less We Understand

We actually have interesting expectations of new technology in this regard. Technology has made it increasingly easy to capture and record the reality that surrounded other people carrying out work. In commercial aviation, the electronic footprint that any flight produces is potentially huge. We can use these data to reconstruct the world as it must have been experienced by other people back then, potentially avoiding the psychologist's fallacy. But capturing such data addresses only one side of the problem. Our ability to make sense of these data, to employ them in a reconstruction of the sensemaking processes of other people at another time and place, has not kept pace with our growing technical ability to register traces of their behavior. In other words, the presumed dominance of human factors in incidents and accidents is not matched by our ability to analyze or understand the human contribution for what it is worth.

Data used in accident analysis often come from a recording of human voices and perhaps other sounds (ruffling charts, turning knobs), which can be coupled to a greater or lesser extent with contemporaneous system or process behavior. A voice trace, however, represents only a partial data record. Human behavior in rich, unfolding settings is much more than the voice trace it leaves behind. The voice trace always points beyond itself, to a world that was unfolding around the practitioners at the time, to tasks, goals, perceptions, intentions, thoughts, and actions that have since evaporated.

But most investigations are formally restricted in how they can couple the cockpit voice recording to the world that was unfolding around the practitioners (e.g., instrument indications, automation-mode settings). In aviation, for example, International Civil Aviation Organization (ICAO Annex 13) prescribes how only those data that can be factually established may be analyzed in the search for cause. This provision often leaves the cockpit voice recording as only a factual, decontextualized, and impoverished footprint of human performance. Making connections between the voice trace and the circumstances and people in which it was grounded quickly falls outside the pale of official analysis and into the realm of what many would call inference or speculation. This inability to make clear connections between behavior and world straightjackets any study of the human contribution to a cognitively noisy, evolving sequence of events. ICAO Annex 13 thus regulates the disembodiment of data: Data must be studied away from their context, for the context and the connections to it are judged as too tentative, too abstract, too unreliable. Such a provision, contradicted by virtually all cognitive psychological research, is devastating to our ability to make sense of puzzling performance.

Apart from the provisions of ICAO Annex 13, this problem is complicated by the fact that current flight-data recorders (FDRs) often do not capture many automation-related traces: precisely those data that are of immediate importance to understanding the problem-solving environment in which most pilots today carry out their work. For example, FDRs in many highly automated aircraft do not record which ground-based navigation beacons were selected by the pilots, which automation-mode control-panel selections on airspeed, heading, altitude, and vertical speed were made, or what was shown on both pilots' moving map displays. As operator work has shifted to the management and supervision of a suite of automated resources, and problems leading to accidents increasingly start in human–machine interactions, this represents a large gap in our ability to access the reasons for human assessments and actions in modern operational workplaces.

INVERTING PERSPECTIVES

Knowing about and guarding against the psychologist's fallacy, against the mixing of realities, is critical to understanding error. When looked at from the position of retrospective outsider, the error can look so very real, so compelling. They failed to notice, they did not know, they should have done this or that. But from the point of view of people inside the situation, as well as potential other observers, this same error is often nothing more than normal work. If we want to begin to understand why it made sense for people to do what they did, we have to reconstruct their local rationality. What did they know? What was their understanding of the situation? What were their multiple goals, resource constraints, pressures? Behavior is rational within situational contexts: People do not come to work to do a bad job. As historian Barbara Tuchman put it: "Every scripture is entitled to be read in the light of the circumstances that brought it forth. To understand the choices open to people of another time, one must limit oneself to what they knew; see the past in its own clothes, as it were, not in ours" (1981, p. 75).

This position turns the exigent social and operational context into the only legitimate interpretive device. This context becomes the constraint on what meaning we, who were not there when it happened, can now give to past controversial assessments and actions. Historians are not the only ones to encourage this switch, this inversion of perspectives, this persuasion to put ourselves in the shoes of other people. In hermeneutics it is known as the difference between *exegesis* (reading out of the text) and *eisegesis* (reading into the text). The point is to read out of the text what is has to offer about *its* time and place, not to read into the text what we want it to say or reveal now. Jens Rasmussen points out that if we cannot find a satisfactory answer to questions such as "how could they not have known?", then this is not because these people were behaving bizarrely. It is because *we* have chosen the wrong frame of reference for understanding their behavior (Vicente, 1999). The frame of reference for understanding people's behavior is their own normal, individual work context, the context they are embedded in and from whose point of view the decisions and assessments made are mostly normal, daily, unremarkable, perhaps even unnoticeable. A challenge is to understand how assessments and actions that from the outside look like errors become neutralized or normalized so that from the inside they appear unremarkable, routine, normal.

If we want to understand why people did what they did, then the adequacy of the insider's representation of the situation cannot be called into question. The reason is that there are no objective features in the domain on which we can base such a judgment. In fact, as soon as we make such a judgment, we have imported criteria from the outside—from another time and place, from another rationality. Ethnographers have always championed the point of view of the person on the inside. Like Rasmussen, Emerson advised that, instead of using criteria from outside the setting to examine mistake and error, we should investigate and apply local notions of competent performance that are honored and used in particular social settings (Vaughan, 1999). This excludes generic rules and motherhoods (e.g., "pilots should be immune to commercial pressures"). Such putative standards ignore the subtle dynamics of localized skills and priority setting, and run roughshod over what would be considered "good" or "competent" or "normal" from inside actual situations. Indeed, such criteria impose a rationality from the outside, impressing a frame of context-insensitive, idealized concepts of practice upon a setting where locally tailored and subtly adjusted criteria rule instead.

The ethnographic distinction between etic and emic perspectives was coined in the 1950s to capture the difference between how insiders view a setting and how outsiders view it. *Emic* originally referred to the language and categories used by people in the culture studied, whereas *etic* language and categories were those of outsiders (e.g., the ethnographer) based on their analysis of important distinctions. Today, emic is often understood to be the view of the world from the inside out, that is, how the world looks from the eyes of the person studied. The point of ethnography is to develop an insider's view of what is happening, an inside-out view. Etic is contrasted as the perspective from the outside in, where researchers or observers attempt to gain access to some portions of an insider's knowledge through psychological methods such as surveys or laboratory studies.

Emic research considers the meaning-making activities of individual minds. It studies the multiple realities that people construct from their experiences with their empirical reality. It assumes that there is no direct access to a single, stable, and fully knowable external reality. Nobody has this access. Instead, all understanding of reality is contextually embedded and limited by the local rationality of the observer. Emic research points at the unique experience of each human, suggesting that any observer's way of making sense of the world is as valid as any other, and that there are no objective criteria by which this sensemaking can be judged correct or incorrect. Emic researchers resist distinguishing between objective features of a situation, and subjective ones. Such a distinction distracts the observer from the situation as it looked to the person on the inside, and in fact distorts this insider perspective.

A fundamental concern is to capture and describe the point of view of people inside a system or situation, to make explicit that which insiders take for granted, see as common sense, find unremarkable or normal. When we want to understand error, we have to embrace ontological relativity not out of philosophical intransigence or philanthropy, but for trying to get the inside-out view. We have to do this for the sake of learning what makes a system safe or brittle. As we saw in chapter 2, for example, the notion of what constitutes an incident (i.e., what is worthy of reporting as a safety threat) is socially constructed, shaped by history, institutional constraints, cultural and linguistic notions. It is negotiated among insiders in the system. None of the structural measures an organization takes to put an incident reporting system in place will have any effect if insiders do not see safety threats as incidents that are worth sending into the reporting system. Nor will the organization ever really improve reporting rates if it does not understand the notion of incident (and, conversely, the notion of normal practice) from the point of view of the people who do it everyday.

To succeed at this, outsiders need to take the inside-out look, they need to embrace ontological relativity, as only this can crack the code to system safety and brittleness. All the processes that set complex systems onto their drifting paths toward failure—the conversion of signals of danger into normal, expected problems, the incrementalist borrowing from safety, the assumption that past operational success is a guarantee of future safety—are sustained through implicit social-organizational consensus, driven by insider language and rationalizations. The internal workings of these processes are simply impervious to outside inspection, and thereby numb to external pressure for change. Outside observers cannot attain an emic perspective, nor can they study the multiple rationalities created by people on the inside if they keep seeing errors and violations. Outsiders can perhaps get some short-term leverage by (re)imposing context-insensitive rules, regulations, or exhortations and making moral appeals for people to follow them, but the effects are generally short-lived. Such measures cannot be supported by operational ecologies. There, actual practice is always under pressure to adapt in an open system, exposed to pressures of scarcity and competition. It will once again inevitably drift into niches that generate greater operational returns at no apparent cost to safety.

ERROR AND (IR)RATIONALITY

Understanding error against the background of local rationality, or rationality for that matter, has not been an automatic by-product of studying the psychology of error. In fact, research into human error had a very rationalist bias up to the 1970s (Reason, 1990), and in some quarters in psychology and human factors such rationalist partiality has never quite disappeared. *Rationalist* means that mental processes can be understood with reference to normative theories that describe optimal strategies. Strategies may be optimal when the decision maker has perfect, exhaustive access to all relevant information, takes time enough to consider it all, and applies clearly defined goals and preferences to making the final choice. In such cases, errors are explained by reference to deviations from this rational norm, this ideal. If the decision turns out wrong it may be because the decision maker did not take enough time to consider all information, or that he or she did not generate an exhaustive set of choice alternatives to pick from. Errors, in other words, are deviant. They are departures from a standard. Errors are irrational in the sense that they require a motivational (as opposed to cognitive) component in their explanation. If people did not take enough time to consider all information, it is because they could not be bothered to. They did not try hard enough, and they should try harder next time, perhaps with the help of some training or procedural guidance. Investigative practice in human factors is still rife with such rationalist reflexes.

It did not take long for cognitive psychologists to find out how humans could not or should not even behave like perfectly rational decision makers. Whereas economists clung to the normative assumptions of decision making (decision makers have perfect and exhaustive access to information for their decisions, as well as clearly defined preferences and goals about what they want to achieve), psychology, with the help of artificial intelligence, posited that there is no such thing as perfect rationality (i.e., full knowledge of all relevant information, possible outcomes, relevant goals), because there is not a single cognitive system in the world (neither human nor machine) that has sufficient computational capacity to deal with it all. Rationality is bounded. Psychology subsequently started to chart people's imperfect, bounded, or local rationality. Reasoning, it discovered, is governed by people's local understanding, by their focus of attention, goals, and knowledge, rather than some global ideal. Human performance is embedded in, and systematically connected to, the situation in which it takes place: It can be understood (i.e., makes sense) with reference to that situational context, not by reference to some universal standard. Human actions and assessments can be described meaningfully only in reference to the localized setting in which they are made; they can be understood by intimately linking them to details of the context that produced and accompanied them. Such research has given *rationality* an interpretive flexibility: What is locally rational does not need to be globally rational. If a decision is locally rational, it makes sense from the point of view of the decision maker, which is what matters if we want to learn about the underlying reasons for what from the outside looks like error. The notion of local rationality removes the need to rely on irrational explanations of error. Errors make sense: They are rational, if only locally so, when seen from the inside of the situation in which they were made.

But psychologists themselves often have trouble with this. They keep on discovering biases and aberrations in decision making (e.g., groupthink, confirmation bias, routine violations) that seem hardly rational, even from within a situational context. These deviant phenomena require motivational explanations. They call for motivational solutions. People should be motivated to do the right thing, to pay attention, to double check. If they do not, then they should be reminded that it is their duty, their job. Notice how easily we slip back into prehistoric behaviorism: Through a modernist system of rewards and punishments (job incentives, bonuses, threats of retribution) we hope to mold human performance after supposedly fixed features of the world.

That psychologists continue to insist on branding such action as irrational, referring it back to some motivational component, may be due to the limits of the conceptual language and power of the discipline. Putatively motivational issues (such as deliberately breaking rules) must themselves be put back into context, to see how human goals (getting the job done fast by not following all the rules to the letter) are made congruent with system goals through a collective of subtle pressures, subliminal messages about organizational preferences, and empirical success of operating outside existing rules. The system wants fast turnaround times, maximization of capacity utilization, efficiency. Given those system goals (which are often kept implicit), rulebreaking is not a motivational shortcoming, but rather an indication of a well-motivated human operator: Personal goals and system goals are harmonized, which in turn can lead to total system goal displacement:

Efficiency is traded off against safety. But psychology often keeps seeing motivational shortcomings. And human factors keeps suggesting counter-measures such as injunctions to follow the rules, better training, or more top-down task analysis. Human factors has trouble incorporating the subtle but powerful influences of organizational environments, structures, processes, and tasks into accounts of individual cognitive practices. In this regard the discipline is conceptually underdeveloped. Indeed, how unstated cultural norms and values travel from the institutional, organizational level to express themselves in individual assessments and actions (and vice versa) is a concern central to sociology, not human factors. Bridging this macro–micro connection in the systematic production of rule violations means understanding the dynamic interrelationships between issues as wide ranging as organizational characteristics and preferences, its environment and history; incrementalism in trading safety off against production, unintentional structural secrecy that fragments problem-solving activities across different groups and departments; patterns and representations of safety-related information that are used as imperfect input to organizational decision making, the influence of hierarchies and bureaucratic accountability on people's choice, and others (e.g., Vaughan, 1996, 1999). The structuralist lexicon of human factors and system safety today has no words for many of these concepts, let alone models for how they go together.

From Decision Making to Sensemaking

In another move away from rationalism and toward the inversion of perspectives (i.e., trying to understand the world the way it looked to the decision maker at the time), large swathes of human factors have embraced the ideas of naturalistic decision making (NDM) over the last decade. By importing cyclical ideas about cognition (situation assessment informs action, which changes the situation, which in turn updates assessment, Neisser, 1976) into a structuralist, normativist psychological lexicon, NDM virtually reinvented decision making (Orasanu & Connolly, 1993). The focus shifted from the actual decision moment, back into the preceding realm of situation assessment.

This shift was accompanied by a methodological reorientation, where decision making and decision makers were increasingly studied in their complex, natural environments. Real decision problems, it quickly turned out, resist the rationalistic format dictated for so long by economics: Options are not enumerated exhaustively, access to information is incomplete at best, and people spend more time assessing and measuring up situations than making decisions—if that is indeed what they do at all (Klein, 1998). In contrast to the prescriptions of the normative model, decision makers tend not to generate and evaluate several courses of action concurrently, in order to then determine the best choice. People do not typically have clear or stable sets of preferences along which they can even rank the enumerated courses of action, picking the best one, nor do most complex decision problems actually have a single correct answer. Rather, decision makers in action tend to generate single options at the time, mentally simulate whether this option would work in practice, and then either act on it, or move on to a new line of thought. NDM also takes the role of expertise more seriously than previous decision-making paradigms: What distinguishes good decision makers from bad decision makers most is their ability to make sense of situations by using a highly organized experience base of relevant knowledge. Once again neatly folding into ideas developed by Neisser, such reasoning about situations is more schema-driven, heuristic, and recognitional than it is computational. The typical naturalistic decision setting does not allow the decision maker enough time or information to generate perfect solutions with perfectly rational calculations. Decision making in action calls for judgments under uncertainty, ambiguity and time pressure. In those settings, options that appear to work are better than perfect options that never get computed.

The same reconstructive, corrective intervention into history that produces our clear perceptions of errors, also generates discrete decisions. What we see as decision making from the outside is action embedded in larger streams of practice, something that flows forth naturally and continually from situation assessment and reassessment. Contextual dynamics are a joint product of how problems in the world are developing and the actions taken to do something about it. Time becomes nonlinear: Decision and action are interleaved rather than temporally segregated. The decision maker is thus seen as in step with the continuously unfolding environment, simultaneously influenced by it and influencing it through his or her next steps. Understanding decision making, then, requires an understanding of the dynamics that lead up to those supposed decision moments, because by the time we get there, the interesting phenomena have evaporated, gotten lost in the noise of action. NDM research is front-loaded: it studies the front end of decision making, rather than the back end. It is interested, indeed, in sensemaking more than in decision making.

Removing *decision making* from the vocabulary of human factors investigations is the logical next step, suggested by Snook (2000). It would be an additional way to avoid counterfactual reasoning and judgmentalism, as decisions that eventually led up to a bad outcome all too quickly become bad decisions:

> Framing such tragedies as decisions immediately focuses our attention on an individual making choices . . . such a framing puts us squarely on a path that leads straight back to the individual decision maker, away from the potentially

powerful contextual features and right back into the jaws of the fundamental attribution error. "Why did they decide . . . ?" quickly becomes "Why did they make the wrong decision?" Hence, the attribution falls squarely onto the shoulders of the decision maker and away from potent situational factors that influence action. Framing the . . . puzzle as a question of meaning rather than deciding shifts the emphasis away from individual decision makers toward a point somewhere "out there" where context and individual action overlap. (Snook, p. 206)

Yet sensemaking is not immune to counterfactual pressure either. If what made sense to the person inside the situation still makes no sense given the outcome, then human factors hastens to point that out (see chap. 5). Even in sensemaking, the normativist bias is an everpresent risk.

THE HINDSIGHT BIAS IS NOT A BIAS
AND IS NOT ABOUT THE PAST

Perhaps the pull in the direction of the position of retrospective outsider is irresistible, inescapable, whether we make lexical adjustments in our investigative repertoire or not. Even with the potentially judgmental notion of decision making removed from the forensic psychological toolbox, it remains incredibly difficult to see the past in its own clothes, not in ours. The fundamental attribution error is alive and well, as Scott Snook puts it (2000, p. 205). We blame the human in the loop and underestimate the influence of context on performance, despite repeated warnings of this frailty in our reasoning. Perhaps we are forever unable to shed our own projection of reality onto the circumstances of people at another time and place. Perhaps we are doomed to digitizing past performance, chunking it up into discrete decision moments that inevitably lure us into counterfactual thinking and judgments of performance instead of explanations. Just as any act of observation changes the observed, our very observations of the past inherently intervene in reality, converting complex histories into more linear, more certain, and disambiguated chronicles. The mechanisms described earlier in this chapter may explain how hindsight influences our treatment of human performance data, but they hardly explain why. They hardly shed light on the energy behind the continual pull toward the position of retrospective outsider; they merely sketch out some of the routes that lead to it.

In order to explain failure, we seek failure. In order to explain missed opportunities and bad choices, we seek flawed analyses, inaccurate perceptions, violated rules—even if these were not thought to be influential or obvious or even flawed at the time (Starbuck & Milliken, 1988). This search for failures is something we cannot seem to escape. It is enshrined in the accident models popular in transportation human factors of our age (see chaps. 1 and 2) and proliferated in the fashionable labels for "human error" that human factors keeps inventing (see chap. 6). Even where we turn away from the etic pitfalls of looking into people's decision making, and focus on a more benign, emic, situated sensemaking, the rationalist, normativist perspective is right around the corner. If we know the outcome was bad, we can no longer objectively look at the behavior leading up to it—it must also have been bad (Fischoff, 1975). To get an idea, think of the Greek mythological figure Oedipus, who shared Jocasta's bed. How large is the difference between Oedipus' memory of that experience before and after he found out that Jocasta was his mother? Once he knew, it was simply impossible for him to look at the experience the same way. What had he missed? Where did he not do his homework? How could he have become so distracted? Outcome knowledge afflicts all retrospective observers, no matter how hard we try not to let it influence us. It seems that bad decisions always have something in common, and that is that they all seemed like a good idea at the time. But try telling that to Oedipus.

The Hindsight Bias Is an Error That Makes Sense, Too

When a phenomenon is so impervious to external pressure to change, one would begin to suspect that it has some adaptive value, that it helps us preserve something, helps us survive. Perhaps the hindsight bias is not a bias, and perhaps it is not about history. Instead, it may be a highly adaptive, forward-looking, rational response to failure. This putative bias may be more about predicting the future than about explaining the past. The linearization and simplification that we see happen in the hindsight bias may be a form of abstraction that allows us to export and project our and others' experiences onto future situations. Future situations can never be predicted at the same level of contextual detail as the new view encourages us to explain past situations. Predictions are possible only because we have created some kind of model for the situation we wish to gain control over, not because we can exhaustively foresee every contextual factor, influence, or data point. This model—any model—is an abstraction away from context, an inherent simplification. The model we create—naturally, effortlessly, automatically—after past events with a bad outcome inevitably becomes a model of binary choices, bifurcations, and unambiguous decision moments. That is the only useful kind of model we can take with us into the future if we want to guard against the same type of pitfalls and forks in the road.

The hindsight bias, then, is about learning, not about explaining. It is forward looking, not backward looking. This applies to ourselves and our own failures as much as it applies to our observations of other people's failures. When confronted by failures that occurred to other people, we may imperatively be tripped into vicarious learning, spurned by our own urge for survival: What do I do to avoid that from happening to me? When confronted by our own performance, we have no privileged insight into our own failures, even if we would like to think we do. The past is the past, whether it is our own or somebody else's. Our observations of the past inevitably intervene and change the observed, no matter whose past it is. This is something that the fundamental attribution error cannot account for. It explains how we overestimate the influence of stable, personal characteristics when we look at other people's failings. We underplay the influence of context or situational factors when others do bad things.

But what about our own failings? Even here we are susceptible to reframing past complexity as simple binary decisions, wrong decisions due to personal shortcomings: things we missed, things we should have done or should not have done. Snook (2000) investigated how, in the fog of post-Gulf War Iraq, two helicopters carrying U.N. peacekeepers were shot down by American fighter jets. The situation in which the shoot-down occurred was full of risk, role ambiguity, operational complexity, resource pressure, slippage between plans and practice. Yet immediately after the incident, all of this gets converted into binary simplicity (a choice to err or not to err) by DUKE—the very command onboard the airborne control center whose job it was not to have such things happen. Allowing the fighters to shoot down the helicopters was their error, yet they do not blame context at all, as the fundamental attribution error predicts they should. It was said of the DUKE that immediately after the incident: "he hoped we had not shot down our own helicopters and that he couldn't believe anybody could make that dumb a mistake" (Snook, p. 205).

It is DUKE himself who blames his own dumb mistake. As with the errors in chapter 3, the dumb mistake is something that jumps into view only with knowledge of outcome, its mistakeness a function of the outcome, its dumbness a function of the severity of the consequences. While doing the work, helping guide the fighters, identifying the targets, all DUKE was doing was his job. It was normal work. He was not sitting there making dumb mistakes. They are a product of hindsight, his own hindsight, directed at his own "mistakes." The fundamental attribution error does not apply. It is overridden.

The fighter pilots, too, engage in self-blame, literally converting the ambiguity, risk, uncertainty, and pressure of their encounter with potentially hostile helicopters into a linear series of decision errors, where they repeatedly and consistently took wrong turns on their road to perdition (we misidentified, we engaged, and we destroyed): "Human error did occur. We misidentified the helicopters; we engaged them; and we destroyed them. It was a tragic and fatal mistake" (Tiger 02 quoted in Snook, 2000, p. 205). Again, the fundamental attribution error makes the wrong prediction. If it were true, then these fighter pilots would tend to blame context for their own errors. Indeed, it was a rich enough context—fuzzy, foggy, dangerous, multi-player, pressurized, risky—with plenty of blameworthy factors to go around, if that is where you would look. Yet these fighter pilots do not. "We" misidentified, "we" engaged, "we" destroyed. The pilots had the choice not to; in fact, they had a series of three choices not to instigate a tragedy. But they did. Human error did occur. Of course, elements of self-identity and control are wrapped up in such an attribution, a self-identity for which fighter pilots may well be poster children.

It is interesting to note that the tendency to convert past complexity into binary simplicity—into twofold choices to identify correctly or incorrectly, to engage or not, to destroy or not—overrides the fundamental attribution error. This confirms the role of the hindsight bias as a catalyst for learning. Learning (or having learned) expresses itself most clearly by doing something differently in the future, by deciding or acting differently, by removing one's link in the accident chain, as fighter pilot Tiger 02 put it: "Remove any one link in the chain and the outcome would be entirely different. I wish to God I could go back and correct my link in this chain—my actions which contributed to this disaster" (Tiger 02, quoted in Snook, 2000, p. 205).

We cannot undo the past. We can only undo the future. But undoing the future becomes possible only when we have abstracted away past failures, when we have decontextualized them, stripped them, cleaned them from the fog and confusion of past contexts, highlighted them, blown them up into obvious choice moments that we, and others, had better get right next time around. Prima facie, the hindsight bias is about misassessing the contributions of past failings to bad outcomes. But if the phenomenon is really as robust as it is documented to be and if it actually manages to override the fundamental attribution error, it is probably the expression of more primary mechanisms running right beneath its surface.

The hindsight bias is a meaningful adaptation. It is not about explaining past failures. It is about preventing future ones. In preparing for future confrontations with situations where we or others might err again, and do not want to, we are in some sense taking refuge from the banality of accidents thesis. The thought that accidents emerge from murky, ambiguous, everyday decision making renders us powerless to do anything meaningful about it. This is where the hindsight bias is so fundamentally adaptive. It highlights for us where we could fix things (or where we think we could fix things), so that the bad thing does not happen again. The hindsight bias is not a bias at all, in the sense of a departure from some rational norm. The hindsight bias is rational. It in itself represents and sustains rationality. We have to see the past as a binary choice, or a linear series of binary choices, because that is the only way we can have any hope of controlling the future. There is no other basis for learning, for adapting. Even if those adaptations may consist of rather coarse adjustments, of undamped and overcontrolling regulations, even if these adaptations occur at the cost of making oversimplified predictions. But making oversimplified predictions of how to control the future is apparently better than having no predictions at all.

Quite in the spirit of Saint Augustine, we accept the reality of errors, and the guilt that comes with it, in the quest for control over our futures. Indeed, the human desire to attain control over the future surely predates the Scientific Revolution. The more refined and empirically testable tools for gaining such control, however, were profoundly influenced and extended by it. Control could best be attained through mechanization and technology—away from nature, spirit, away from primitive incantations to divine powers to spare us the next disaster. These Cartesian–Newtonian reflexes have tumbled down the centuries to proffer human factors legitimate routes for gaining control over complex, dynamic futures today and tomorrow. For example, when we look at the remnants of a crashed automated airliner, we, in hindsight, exclaim, "they should have known they were in open descent mode!" The legitimate solution for meeting such technology surprises is to throw more technology at the problem (additional warning systems, paperless cockpits, automatic cocoons). But more technology often creates more problems of a kind we have a hard time anticipating, rather than just solving existing ones.

As another example, take the error-counting methods discussed in chapter 3. A more formalized way of turning the hindsight bias into an oversimplified forward looking future controller is hardly imaginable. Errors, which are uniquely the product of retrospective observations conducted from the outside, are measured, categorized, and tabulated. This produces bar charts that putatively point toward a future, jutting their dire predictions of rule violations or proficiency errors out into a dark and fearful time to come, away from a presumed "safe" baseline. It is normativism in pretty forms and colors.

> These forecasting techniques, which are merely an assignment of categories and numbers to the future, are appearing everywhere. However, their categorical and numerical output can at best be as adequate or as inadequate as the input. Using such forecasts as a strategic tool is only a belief that numbers are meaningful in relation to the fearful future. Strategy becomes a matter of controlling the future by labelling it, rather than continually re-evaluating the uncertain situation. This approach, searching for the right and numerical label to represent the future, is more akin to numerology or astrology. It is the modern-day ritual equivalent of "reading the runes" or "divining the entrails."
> (Angell & Straub, 1999, p. 184)

Human factors holds on to the belief that numbers are meaningful in relation to a fearful future. And why not? Measuring the present and mathematically modeling it (with barcharts, if you must), and thereby predicting and controlling the future has been a legitimate pursuit since at least the 16th century. But as chapters 1, 2, and 3 show, such pursuits are getting to be deeply problematic. In an increasingly complex, dynamic sociotechnical world, their predictive power is steadily eroding. It is not only a problem of garbage in, garbage out (the categorical and numerical output is as adequate or inadequate as the input). Rather, it is the problem of not seeing that we face an uncertain situation in the first place, where mistake, failure, and disaster are incubated in systems much larger, much less transparent, and much less deterministic than the counters of individual errors believe.

This, of course, is where the hindsight bias remains a bias. But it is a bias about the future, not about the past. We are biased to believe that thinking about action in terms of binary choices will help us undo bad futures, that it prepares us sufficiently for coming complexity. It does not. Recall how David Woods (2003) put it: Although the past is incredible (DUKE couldn't believe anybody could make that dumb a mistake), the future is implausible. Mapping digitized historic lessons of failure (which span the arc from error bar charts to Tiger 02's wish to undo his link in the chain) into the future will only be partly effective. Stochastic variation and complexity easily outrun our computational capacity to predict with any accuracy.

PRESERVATION OF SELF- AND SYSTEM IDENTITY

There is an additional sense in which our dealings with past failures go beyond merely understanding what went wrong and preventing recurrence. Mishaps are surprising relative to prevailing beliefs and assumptions about the system in which they happen, and investigations are inevitably affected by the concern to reconcile a disruptive event with existing views and beliefs about the system. Such reconciliation is adaptive too. Our reactions to failure, and our investigations into failure, must be understood against the backdrop of the "fundamental surprise error" (Lanir, 1986) and examined for the role they play in it. Accidents tend to create a profound asymmetry between our beliefs (or hopes) of a basically safe system, and new evidence that may suggest that it is not. This is the fundamental surprise: the astonishment that we feel when the most basic assumptions we held true about the world may turn out to be untrue. The asymmetry creates a tension, and this tension creates pressure for change: Something will have to give. Either the belief needs changing (i.e., we have to acknowledge that the system is not basically safe—that mistake, mishap, and disaster are systematically organized by that system; Vaughan, 1996), or we change the people involved in the mishap—even if this means us. We turn them into unrepresentative, uniquely bad individuals:

> The pilots of a large military helicopter that crashed on a hillside in Scotland in 1994 were found guilty of gross negligence. The pilots did not survive—29 people died in total—so their side of the story could never be heard. The official inquiry had no problems with "destroying the reputation of two good men," as a fellow pilot put it. Potentially fundamental vulnerabilities (such as 160 reported cases of Uncommanded Flying Control Movement or UFCM in computerized helicopters alone since 1994) were not looked into seriously. (Dekker, 2002, p. 25)

When we elect to "destroy the reputation of two good men," we have committed the fundamental surprise error. We have replaced a fundamental challenge to our assumptions, our beliefs (the fundamental surprise) with a mere local one: The pilots were not as good as we thought they were, or as good as they should have been. From astonishment (and its concomitant: fear about the basic safety of the system, as would be raised by 160 cases of UFCM) we move to mere, local wonder: How could they not have seen the hill? They must not have been very good pilots after all. Thus we strive to preserve our self- and system identity. We pursue an adaptive strategy of safeguarding the essence of our world as we understand it. By letting the reputation of individual decision makers slip, we have relieved the tension between broken beliefs (the system is not safe after all) and fervent hopes that it still is. That phenomena such as the hindsight bias and the fundamental attribution error may not be primary, but rather ancillary expressions of more adaptive, locally rational, and useful identity-preserving strategies for the ones committing them, is consonant with observations of a range of reasoning errors. People keep committing them not because they are logical (i.e., globally rational) or because they only produce desired effects, but because they serve an even weightier purpose: "This dynamic, this 'striving to preserve identity,' however strange the means or effects of such striving, was recognised in psychiatry long ago. [This phenomenon] is seen not as primary, but as attempts (however misguided) at restitution, at reconstructing a world reduced by complete chaos" (Sacks, 1998, p. 7).

However "strange the means or effects of such striving," the fundamental surprise error allows us to rebuild a world reduced by chaos. And the hindsight bias allows us to predict and avoid future roads to perdition. Through the fundamental surprise error, we rehabilitate our faith in something larger than ourselves, something in which we too are vulnerable to breakdown, something that we too are at the mercy of in varying degrees. Breaking out of such locally rational reasoning, where the means and consequences of our striving for preservation and rehabilitation create strange and undesirable side effects (blaming individuals for system failures, not learning from accidents, etc.) requires extraordinary courage. It is not very common.

Yet people and institutions may not always commit the fundamental surprise error, and may certainly not do so intentionally. In fact, in the immediate aftermath of failure, people may be willing to question their underlying assumptions about the system they use or operate. Perhaps things are not as safe as previously thought; perhaps the system contains vulnerabilities and residual weaknesses that could have spawned this kind of failure earlier, or worse, could do it again. Yet such openness does not typically last long. As the shock of an accident subsides, parts of the system mobilize to contain systemic self-doubt and change the fundamental surprise into a merely local hiccup that temporarily ruffled an otherwise smooth operation. The reassurance is that the system is basically safe. It is only some people or other parts in it that are unreliable.

In the end, it is not often that an existing view of a system gives in to the reality of failure. Instead, to redress the asymmetry, the event or the players in it are changed to fit existing assumptions and beliefs about the system, rather than the other way around. Expensive lessons about the system as a whole, and the subtle vulnerabilities it contains, can go completely unlearned. Our inability to deal with the fundamental surprise of failure shines through the investigations we commission. The inability to really learn is sometimes legitimized and institutionalized through resource-intensive official investigations. The cause we end up attributing to an accident may sometimes be no more than the "cause" we can still afford, given not just our financial resources, but also our complex of hopes and beliefs in a safe and fair world. As Perrow (1984) has noted:

> Formal accident investigations usually start with an assumption that the operator must have failed, and if this attribution can be made, that is the end of serious inquiry. Finding that faulty designs were responsible would entail enormous shutdown and retrofitting costs; finding that management was responsible would threaten those in charge, but finding that operators were responsible preserves the system, with some soporific injunctions about better training. (p. 146)

Real change in the wake of failure is often slow to come. Few investigations have the courage to really challenge our beliefs. Many keep feeding the hope that the system is still safe—except for this or that little broken component, or this or that Bad Apple. The lack of courage shines through how we deal with human error, through how we react to failure. It affects the words we choose, the rhetoric we rely on, the pathways for "progress" we put our bets on.

Which cause can we afford? Which cause renders us too uncomfortable? Accuracy is not the dominant criterion, but plausibility is—plausibility from the perspective of those who have to accommodate the surprise that the failure represents for them and their organization, their worldview. "Is it plausible?," is the same as asking, "Can we live (on) with this explanation? Does this explanation help us come to terms with the puzzle of bad performance?" Answering this question, and generating such comfort and self-assurance, is one purpose that our analysis of past failures has to fulfill, even if it becomes a selective oversimplification because of it. Even if, in the words of Karl Weick (1995), they make lousy history.

Chapter 5

If You Lose Situation Awareness, What Replaces It?

The hindsight bias has ways of getting entrenched in human factors thinking. One such way is our vocabulary. "Losing situation awareness" or "deficient situation awareness" have become legitimate characterizations of cases where people did not exactly know where they were or what was going on around them. In many applied as well as some scientific settings, it is acceptable to submit "loss of situation awareness" as an explanation for why people ended up where they should not have ended up, or why they did what they, in hindsight, should not have done. Navigational incidents and accidents in transportation represent one category of cases where the temptation to rely on situation awareness as an elucidatory construct appears irresistible. If people end up where they should not, or where they did not intend to end up, it is easy to see that as a deficient awareness of the cues and indications around them. It is easy to blame a loss of situation awareness.

One such accident happened to the Royal Majesty, a cruise ship that was sailing from Bermuda to Boston in the Summer of 1995. It had more than 1,000 people onboard. Instead of Boston, the Royal Majesty ended up on a sandbank close to the Massachusetts shore. Without the crew noticing, it had drifted 17 miles off course during a day and a half of sailing (see Fig. 5.1).

Investigators discovered afterward that the ship's autopilot had defaulted to DR (Dead Reckoning) mode (from NAV, or Navigation mode) shortly after departure. DR mode does not compensate for the effects of wind and other drift (waves, currents), which NAV mode does. A northeasterly wind pushed the ship steadily off its course, to the side of its intended track. The U.S. National Transportation Safety Board investigation into the

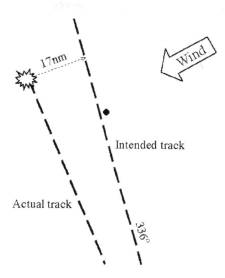

FIG. 5.1. The difference between where the Royal Majesty crew thought they were headed (Boston) and where they actually ended up: a sandbank near Nantucket.

accident judged that "despite repeated indications, the crew failed to recognize numerous opportunities to detect that the vessel had drifted off track" (NTSB, 1995, p. 34).

But "numerous opportunities" to detect the nature of the real situation become clear only in hindsight. With hindsight, once we know the outcome, it becomes easy to pick out exactly those clues and indications that would have shown people where they were actually headed. If only they had focused on this piece of data, or put less confidence in that indication, or had invested just a little more energy in examining this anomaly, then they would have seen that they were going in the wrong direction. In this sense, *situation awareness* is a highly functional or adaptive term for us, for those struggling to come to terms with the rubble of a navigational accident. Situation awareness is a notation that assists us in organizing the evidence available to people at the time, and can provide a starting point for understanding why this evidence was looked at differently, or not at all, by those people. Unfortunately, we hardly ever push ourselves to such understanding. Loss of situation awareness is accepted as sufficient explanation too quickly too often, and in those cases amounts to nothing more than saying human error under a fancy new label.

The kinds of notations that are popular in various parts of the situation-awareness literature are one indication that we quickly stop investigating, researching any further, once we have found human error under that new guise. Venn-type diagrams, for example, can point out the mismatch be-

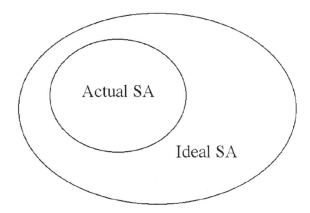

FIG. 5.2. A common, normative characterization of situation awareness. The large circle depicts ideal awareness (or potential awareness) whereas the smaller circle represents actual awareness.

tween actual and ideal situation awareness. They illustrate the difference between what people were aware of in a particular situation, and what they could (or should) ideally have been aware of (see Fig. 5.2). Once we have found a mismatch between what we now know about the situation (the large circle) and what people back then apparently knew about it (the small one), that in itself is explanation enough. They did not know, but they could or should have known. This does not apply only in retrospect, by the way. Even design problems can be clarified through this notation, and performance predictions can be made on its basis. When the aim is designing for situation awareness, the Venn diagram can show what people should pick up in a given setting, versus what they are likely to actually pick up. In both cases, awareness is a relationship between that which is objectively available to people in the outside world on the one hand, and what they take in, or understand about it, on the other. Terms such as *deficient situation awareness* or *loss of situation awareness* confirm human factors' dependence on a kind of subtractive model of awareness. The Venn diagram notation can also be expressed in an equation that reflects this:

$$\text{Loss of SA} = f(\text{large circle} - \text{small circle}) \qquad (1)$$

In Equation 1, "loss of SA" equals "deficient SA" and SA stands for situation awareness. This also reveals the continuing normativist bias in our understanding of human performance. Normativist theories aim to explain mental processes by reference to ideals or normative standards that describe optimal strategies. The large Venn circle is the norm, the standard, the ideal. Situation awareness is explained by reference to that ideal: Actual situation awareness is a subtraction from that ideal, a shortfall, a deficit, indeed, a "loss." Equation 2 becomes:

$$\text{Loss of SA} = f(\text{what I know now} - \text{what you knew then}) \qquad (2)$$

Loss of situation awareness, in other words, is the difference between what I know about a situation now (especially the bits highlighted by hindsight) and what somebody else apparently knew about that situation then. Interestingly, situation awareness is nothing by itself, then. It can only be expressed as a relative, normativist function, for example, the difference between what people apparently knew back then versus what they could or should have known (or what we know now). Other than highlighting the mismatch between the large and the little circle in a Venn diagram, other than revealing the elements that were not seen but could or should have been seen and understood, there is little there. Discourse and research around situation awareness may so far have shed little light on the actual processes of attentional dynamics. Rather, situation awareness has given us a new normativist lexicon that provides large, comprehensive notations for perplexing data (How could they not have noticed? Well, they lost SA). Situation awareness has legitimized the proliferation of the hindsight bias under the pretext of adding to the knowledge base. None of this may really help our understanding of awareness in complex dynamic situations, instead truncating deeper research, and shortchanging real insight.

THE MIND–MATTER PROBLEM

Discourse about situation awareness is a modern installment of an ancient debate in philosophy and psychology, about the relationship between matter and mind. Surely one of the most vexing problems, the coupling between matter and mind, has occupied centuries of thinkers. How is it that we get data from the outside world inside our minds? What are the processes by which this happens? And how can the products of these processes be so divergent (I see other things than you, or other things than I saw yesterday)? All psychological theories, including those of situation awareness, implicitly or explicitly choose a position relative to the mind–matter problem.

Virtually all theories of situation awareness rely on the idea of correspondence—a match, or correlation, between an external world of stimuli (elements) and an internal world of mental representations (which gives meaning to those stimuli). The relationship between matter and mind, in other words, is one of letting the mind create a mirror, a mental simile, of matter on the outside. This allows a further elaboration of the Venn diagram notation: Instead of "ideal" versus "actual" situation awareness, the captions of the circles in the diagram could read "matter" (the large circle) and "mind" (the little circle). Situation awareness is the difference between what is out there in the material world (matter) and what the observer sees or understands (mind). Equation 1 can be rewritten as Equation 3:

$$\text{Loss of SA} = f(\text{matter} - \text{mind}) \tag{3}$$

Equation 3 describes how a loss of situation awareness, or deficient situation awareness, is a function of whatever was in the mind subtracted from what was available matter. The portion of matter that did not make it into mind is lost, it is deficient awareness. Such thinking, of course, is profoundly Cartesian. It separates mind from matter as if they are distinct entities: a *res cogitans* and a *res extensa*. Both exist as separate essentials of our universe, and one serves as the echo, or imitate, of the other.

One problem of such dualism lies, of course, in the assumptions it makes. An important assumption is what Feyerabend called the "autonomy principle" (see chap. 3): that facts exist in some objective world, equally accessible to all observers. The autonomy principle is what allows researchers to draw the large circle of the Venn diagram: It consists of matter available to, as well as independent from, any observer, whose awareness of that matter takes the form of an internal simile. This assumption is heavily contested by radical empiricists. Is there really a separation between *res extensa* and *res cogitans*? Can we look at matter as something "out there," as something independent of (the minds of) observers, as something that is open to enter the awareness of anyone?

If the autonomy principle is right, then deficient situation awareness is a result of the actual SA Venn circle being too small, or misdirected relative to the large (ideal SA) Venn circle. But if you lose situation awareness, what replaces it? No theories of cognition today can easily account for a mental vacuum, for empty-headedness. Rather, people always form an understanding of the situation unfolding around them, even if this understanding can, in hindsight, be shown to have diverged from the actual state of affairs. This does not mean that this mismatch has any relevance in explaining human performance at the time. For people doing work in a situated context, there is seldom a mismatch, if ever. Performance is driven by the desire to construct plausible, coherent accounts, a good story of what is going on. Weick (1995) reminded us that such sensemaking is not about accuracy, about achieving an accurate mapping between some objective, outside world and an inner representation of that world. What matters to people is not to produce a precise internalized simile of an outside situation, but to account for their sensory experiences in a way that supports action and goal achievement. This converts the challenge of understanding situation awareness.

From studying the mapping accuracy between an external and internal world, it requires the investigation of why people thought they were in the right place, or had the right assessment of the situation around them. What made that so? The adequacy or accuracy of an insider's representation of the situation cannot be called into question: It is what counts for him or her, and it is that which drives further action in that situation. The internal, subjective world is the only one that exists. If there is an objective, external reality, we could not know it.

Getting Lost at the Airport

Let us now turn to a simple case, to see how these things play out. Runway incursions (aircraft taxiing onto runways for which they did not have a clearance) are an acute category of such cases in transportation today. Runway incursions are seen as a serious and growing safety problem worldwide, especially at large, controlled airports (where air-traffic control organizes and directs traffic movements). Hundreds of incursions occur every year, some leading to fatal accidents. Apart from straying onto a runway without a clearance, the risk of colliding with something else at an airport is considerable. Airports are tight and dynamic concentrations of cars, buses, carts, people, trucks, trucks plus aircraft, and aircraft, all moving at speeds varying from a few knots to hundreds of knots. (And then fog can settle over all of that). The number of things to hit is much larger on the ground than it is in the air, and the proximity to those things is much closer. And because of the layout of taxiways and ramps, navigating an aircraft across an airport can be considerably more difficult than navigating it in flight.

When runway incursions occur, it can be tempting to blame a loss of situation awareness. Here is one such case, not a runway incursion, but a taxiway incursion. This case is illustrative not only because it is relatively simple, but also because all regulations had been followed in the design and layout of the airport. Safety cases had been conducted for the airport, and it had been certified as compliant with all relevant rules. Incidents in such an otherwise safe system can happen even when everybody follows the rules. This incident happened at Stockholm Arlanda (the international airport) in October 2002. A Boeing 737 had landed on runway 26 (the opposite of runway 08, which can be seen in Fig. 5.3) and was directed by air traffic control to taxi to its gate via taxiways ZN and Z (called "Zulu November" and "Zulu," in aviation speak). The reason for taking ZN was that a tow truck with an aircraft was coming from the other direction. It had been cleared to use ZP (Zulu Papa) and then to turn right onto taxiway X (X-ray). But the 737 did not take ZN. To the horror of the tow truck driver, it carried on following ZP instead, almost straight into the tow truck. The pilots saw the truck in time, however, and managed to stop. The truck driver had to push his air-

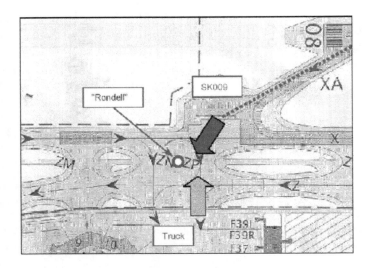

FIG. 5.3.   The near-collision between a Boeing 737 and tow truck. The 737
came off the runway after landing, and had been cleared to take ZN. It took
ZP instead, almost running into the tow truck (which was pulling another aircraft across ZP; Statens Haverikommision, 2003).

craft backward in order to clear up the jam. Did the pilots lose situation
awareness? Was their situation awareness deficient? There were signs pointing out where taxiway ZN ran, and those could be seen from the cockpit.
Why did the crew not take these cues into account when coming off the
runway?

Such questions consistently pull us toward the position of retrospective
outsider, looking down onto the developing situation from a God's-eye
point of view. From there we can see the mismatch grow between where
people were and where they thought they were. From there we can easily
draw the circles of the Venn diagram, pointing out a deficiency or a shortcoming in the awareness of the people in question. But none of that explains much. The mystery of the matter–mind problem is not going to go
away just because we say that other people did not see what we now know
they should have seen. The challenge is to try to understand why the crew
of the 737 thought that they were right—that they were doing exactly what
air-traffic control had told them to do: follow taxiway ZN to Z. The commitment of an antidualist position is to try to see the world through the eyes of
the protagonists, as there is no other valid perspective. The challenge with
navigational incidents, indeed, is not to point out that people were not in
the spot they thought they were, but to explain why they thought they were
right. The challenge is to begin to understand on the basis of what cues
people (thought) they knew where they were.

The first clue can be found in the response of the 737 crew after they had been reminded by the tower to follow ZN (they had now stopped, facing the tow truck head-on). "Yeah, it's the chart here that's a little strange," said one of the pilots (Statens Haverikommision, 2003, p. 8). If there was a mismatch, it was not between the actual world and the crew's model of that world. Rather, there was a mismatch between the chart in the cockpit and the actual airport layout. As can be seen in Fig. 5.3, the taxiway layout contained a little island, or roundabout, between taxiways Zulu and X-ray. ZN and ZP were the little bits going between X-ray to Zulu, around the roundabout. But the chart available in the cockpit had no little island on it. It showed no roundabout (see Fig. 5.4).

Even here, no rules had been broken. The airport layout had recently changed (with the addition of the little roundabout) in connection with the construction of a new terminal pier. It takes time for the various charts to be updated, and this simply had not happened yet at the company of the crew in question. Still, how could the crew of the 737 have ended up on the wrong side of that area (ZP instead of ZN), whether there was an island shown on their charts or not? Figure 5.5 contains more clues. It shows the roundabout from the height of a car (which is lower than a 737 cockpit, but from the same direction as an aircraft coming off of runway 26). The crew in question went left of the roundabout, where it should have gone right. The roundabout is covered in snow, which makes it inseparable from the other (real) islands separating taxiways Zulu and X-ray. These other islands consist of grass, whereas the roundabout, with a diameter of about 20 meters, is the same tarmac as that of the taxiways. Shuffle snow onto all of them, however, and they look indistinguishable. The roundabout is no

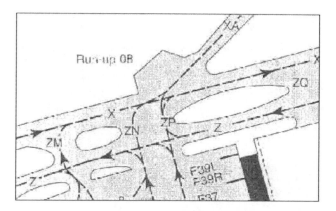

FIG. 5.4. The chart available in the Boeing 737 cockpit at the time. It shows no little roundabout, or island, separating taxiways ZN from ZP (Statens Haverikommision, 2003).

FIG. 5.5.   The roundabout as seen coming off runway 26. The Boeing 737
went left around the roundabout, instead of right (Statens Haverikommision,
2003).

longer a circle painted on the tarmac: It is an island like all others. Without
a roundabout on the cockpit chart, there is only one plausible explanation
for what the island in Fig. 5.5 ahead of the aircraft is: It must be the grassy island to the right of taxiway ZN. In other words, the crew knew where they
were, based on the cues and indications available to them, and based on
what these cues plausibly added up to.

The signage, even though it breaks no rules either, does not help. Taxiway signs are among the most confusing directors in the world of aviation,
and they are terribly hard to turn into a reasonable representation of the
taxiway system they are supposed to help people navigate on. The sign visible from the direction of the Boeing 737 is enlarged in Fig. 5.6. The black
part of the sign is the position part (this indicates what taxiway it is), and the
yellow part is the direction part: This taxiway (ZN) will lead to taxiway Z,
which happens to run at about a right angle across ZN. These signs are
placed to the left of the taxiway they belong to. In other words, the ZN taxiway is on the right of the sign, not on the left. But put the sign in the context of Fig. 5.5, and things become more ambiguous. The black ZN part is

FIG. 5.6.   The sign on the roundabout that is visible for aircraft coming off
runway 26 (Statens Haverikommision, 2003).

now leaning toward the left side of the roundabout, not the right side. Yet the ZN part belongs to the piece of tarmac on the right of the roundabout. The crew never saw as such. For them, given their chart, the roundabout was the island to the right of ZN.

Why not swap the black ZN and the yellow Z parts? Current rules for taxiway signage will not allow it (rules can indeed stifle innovation and investments in safety). And not all airports comply to this religiously either. There may be exceptions when there is no room or when visibility of the sign would be obstructed if placed on the left side. To make things worse, regulations state that taxiway signs leading to a runway need to be placed on both sides of the taxiway. In those cases, the black parts of the signs are often actually adjacent to the taxiway, and not removed from it, as in Fig. 5.5. Against the background of such ambiguity, few pilots actually know or remember that taxiway signs are supposed to be on the left side of the taxiway they belong to. In fact, very little time in pilot training is used to get pilots to learn to navigate around airports, if any. It is a peripheral activity, a small portion of mundane, pedestrian work that merely leads up to and concludes the real work: flying from A to B. When rolling off the runway, and going to taxiway Zulu and then to their gate, this crew knew where they were. Their indications (cockpit chart, snow-covered island, taxiway sign) compiled into a plausible story: ZN, their assigned route, was the one to the left of the island, and that was the one they were going to take. Until they ran into a tow truck, that is. But nobody in this case lost situation awareness. The pilots lost nothing. Based on the combination of cues and indications observable by them at the time, they had a plausible story of where they were. Even if a mismatch can be shown between how the pilots saw their situation and how retrospective, outside observers now see that situation, this has no bearing on understanding how the pilots made sense of their world at the time.

Seeing situation awareness as a measure of the accuracy of correspondence between some outer world and an inner representation carries with it a number of irresolvable problems that have always been connected to such a dualist position. Taking the mind–matter problem apart by separating the two means that the theory needs to connect the two again. Theories of situation awareness typically rely on a combination of two schools in psychological thought to reconstitute this tie, to make this bridge. One is empiricism, a traditional school in psychology that makes claims on how knowledge is chiefly, if not uniquely, based on experience. The second is the information-processing school in cognitive psychology, still popular in large areas of human factors. None of these systems of thought, however, are particularly successful in solving the really hard questions about situation awareness, and may in fact be misleading in certain respects. We look at both of them in turn here. Once that is done, the chapter briefly develops the counterposition on the mind–matter question: an antidualist one (as related to situation awareness). This position is worked out further in the rest of this chapter, using the Royal Majesty case as an example.

EMPIRICISM AND THE PERCEPTION OF ELEMENTS

Most theories of situation awareness actually leave the processes by which matter makes it into mind to the imagination. A common denominator, however, appears to be the perception of elements in the environment. Elements are the starting point of perceptual and meaning-making processes. It is on the basis of these elements that we gradually build up an understanding of the situation, by processing such elementary stimulus information through multiple stages of consciousness or awareness ("levels of SA"). Theories of situation awareness borrow from empiricism (particularly British empiricism), which assumes that the organized character and the meaningfulness of our perceptual world are achieved by matching incoming stimuli with prior experience through a process called *association*. In other words, the world as we experience it is disjointed (consisting of elements) except when mediated by previously stored knowledge. Correspondence between mind and matter is made by linking incoming impressions through earlier associations.

Empiricism in its pure form is nothing more than saying that the major source of knowledge is experience, that we do not know about the world except through making contact with that world with our sensory organs. Among Greek philosophers around the 5th century B.C., empiricism was accepted as a guide to epistemology, as a way of understanding the origin of knowledge. Questions already arose, however, on whether all psychic life could be reduced to sensations. Did the mind have a role to play at all in turning perceptual impressions into meaningful percepts? The studies of perception by Johannes Kepler (1571–1630) would come to suggest that the mind had a major role, even though he himself left the implications of his findings up to other theoreticians. Studying the eyeball, Kepler found that it actually projects an inverted image on the retina at the back. Descartes, himself dissecting the eye of a bull to see what image it would produce, saw the same thing. If the eye inverts the world, how can we see it the right way up? There was no choice but to appeal to mental processing. Not only is the image inverted, it is also two-dimensional, and it is cast onto the backs of two eyes, not one. How does all that get reconciled in a single coherent, upright percept? The experiments boosted the notion of impoverished, meaning-deprived stimuli entering our sensory organs, in need of some serious mental processing work from there on. Further credence to the perception of elements was given by the 19th-century discovery of photoreceptors in the human eye. This mosaic of retinal receptors appeared to chunk up any visual percept coming into the eyeball. The resulting fragmented neural signals had to be sent up the perceptual pathway in the brain for further processing and scene restoration.

British empiricists such as John Locke (1632–1704) and George Berkeley (1685–1753), though not privy to 19th-century findings, were confronted with the same epistemological problems that their Greek predecessors had struggled with. Rather than knowledge being innate, or the chief result of reasoning (as claimed by rationalists of that time), what role did experience have in creating knowledge? Berkeley, for example, wrestled with the problem of depth perception (not a negligible problem when it comes to situation awareness). How do we know where we are in space, in relation to objects around us? Distance perception, to Berkeley, though created through experience, was itself not an immediate experience. Rather, distance and depth are additional aspects of visual data that we learn about through combinations of visual, auditory, and tactile experiences. We understand distance and depth in current scenes by associating incoming visual data with these earlier experiences. Berkeley reduced the problem of space perception to more primitive psychological experiences, decomposing the perception of distance and magnitude into constituent perceptual elements and processes (e.g., lenticular accommodation, blurring of focus). Such deconstruction of complex, intertwined psychological processes into elementary stimuli turned out to be a useful tactic. It encouraged many after him, among them Wilhelm Wundt and latter-day situation-awareness theorists, to analyze other experiences in terms of elements as well. Interestingly, neither all prehistoric empiricists nor all British empiricists could be called dualists in the same way that situation-awareness theorists can be. Protagoras, a contemporary of Plato around 430 B.C., had already said that "man is the measure of all things." An individual's perception is true to him, and cannot be proven untrue (or inferior or superior) by some other individual. Today's theories of situation awareness, with their emphasis on the accuracy of the mapping between matter and mind, are very much into inferiority and superiority (deficient SA vs. good SA) as something that can be objectively judged. This would not have worked for some of the British empiricists either.

To Berkeley, who disagreed with earlier characterizations of an inner versus an outer world, people can actually never know anything but their experiences. The world is a plausible but unproved hypothesis. In fact, it is a fundamentally untestable hypothesis, because we can only know our own experience. Like Protagoras before him, Berkeley would not have put much stock in claims of the possibility of superior or ideal situation awareness, as such a thing is logically impossible. There are no superlatives when it comes to knowledge through experience. For Berkeley too, this meant that even if there is an objective world out there (the large circle in the Venn diagram), we could never know it. It also meant that any characterization of such an objective world with the aim of understanding somebody's perception, somebody's situation awareness, would have been nonsense.

Experiments, Empiricism, and Situation Awareness

Wilhelm Wundt is credited with founding the first psychological laboratory in the world at the University of Leipzig in the late 1870s. The aim of his laboratory was to study mental functioning by deconstructing it into separate elements. These could then be combined to understand perceptions, ideas, and other associations. Wundt's argument was simple and compelling, and versions of it are still used in psychological method debates today. Although the empirical method had been developing all around psychology, it was still occupied with grand questions of consciousness, soul, and destiny, and it tried to gain access to these issues through introspection and rationalism. Wundt argued that these were questions that should perhaps be asked at the logical end point of psychology, but not at the beginning. Psychology should learn to crawl before trying to walk. This justified the appeal of the elementarist approach: chopping the mind and its stimuli up into minute components, and studying them one by one. But how to study them?

Centuries before, Descartes had argued that mind and matter not only were entirely separate, but should be studied using different methods as well. Matter should be investigated using methods from natural science (i.e., the experiment), whereas mind should be examined through processes of meditation, or introspection. Wundt did both. In fact, he combined the natural science tradition with the introspective one, molding them into a novel brand of psychological experimentation that still governs much of human factors research to this day. Relying on complicated sets of stimuli, Wundt investigated sensation and perception, attention, feeling, and association. Using intricate measurements of reaction times, the Leipzig laboratory hoped they would one day be able to achieve a chronometry of mind (which was not long thereafter dismissed as infeasible).

Rather than just counting on quantitative experimental outcomes, Wundt asked his subjects to engage in introspection, to reflect on what had happened inside their minds during the trials. Wundt's introspection was significantly more evolved and demanding than the experimental "report" psychologists ask their subjects for today. Introspection was a skill that required serious preparation and expertise, because the criteria for gaining successful access to the elementary makeup of mind were set very high. As a result, Wundt mostly used his assistants. Realizing that the contents of awareness are in constant flux, Wundt produced rigid rules for the proper application of introspection: (a) The observer, if at all possible, must be in a position to determine when the process is to be introduced; (b) he or she must be in a state of "strained attention"; (c) the observation must be capable of being repeated several times; (d) the conditions of the experiment must be such that they are capable of variation through introduction or elimination of certain stimuli and through variation of the strength and quality of the stimuli.

Wundt thus imposed experimental rigor and control on introspection. Similar introspective rigor, though different in some details and prescriptions, is applied in various methods for studying situation awareness today. Some techniques involve blanking or freezing of displays, with researchers then going in to elicit what participants remember about the scene. This requires active introspection. Wundt would have been fascinated, and he probably would have had a thing or two to say about the experimental protocol. If subjects are not allowed to say when the blanking or freezing is to be introduced, for example (Wundt's first rule), how does that compromise their ability to introspect? In fact, the blanking of displays and handing out a situation-awareness questionnaire is more akin to the Würzburg school of experimental psychology that started to compete with Wundt in the late 19th century. The Würzburgers pursued "systematic experimental introspection" by having subjects pursue complex tasks that involved thinking, judging, and remembering. They would then have their subjects render a retrospective report of their experiences during the original operation. The whole experience had to be described time period by time period, thus chunking it up. In contrast to Wundt, and like situation-awareness research participants today, Würzburg subjects did not know in advance what they were going to have to introspect.

Others today disagree with Descartes' original exhortation and remain fearful of the subjectivist nature of introspection. They favor the use of clever scenarios in which the outcome, or behavioral performance of people, will reveal what they understood the situation to be. This is claimed to be more of a natural science approach that stays away from the need to introspect. It relies on objective performance indicators instead. Such an approach to studying situation awareness could be construed as neobehaviorist, as it equates the study of behavior with the study of consciousness. Mental states are not themselves the object of investigation: Performance is. If desired, such performance can faintly hint at the contents of mind (situation awareness). But that itself is not the aim; it cannot be, because through such pursuits psychology (and human factors) would descend into subjectivism and ridicule. Watson, the great proponent of behaviorism, would himself have argued along these lines.

Additional arguments in favor of performance-oriented approaches include the assertation that introspection cannot possibly test the contents of awareness, as it necessarily appeals to a situation or stimulus from the past. The situation on which people are asked to reflect has already disappeared. Introspection thus simply probes people's memory. Indeed, if you want to study situation awareness, how can you take away the "situation" by blanking or freezing their world, and still hope you have relevant "awareness" left to investigate by introspection? Wundt, as well as many of today's situation awareness researchers, may in part have been studying memory, rather than the contents of consciousness.

Wundt was, and still remains, one of the chief representatives of the elementarist orientation, pioneered by Berkeley centuries before, and perpetuated in modern theories of situation awareness. But if we perceive elements, if the eyeball deals in two-dimensional, fragmented, inverted, meaning-deprived stimuli, then how does order in our perceptual experience come about? What theory can account for our ability to see coherent scenes, objects? The empiricist answer of association is one way of achieving such order, of creating such interelementary connections and meaning. Order is an end product, it is the output of mental or cognitive work. This is also the essence of information processing, the school of thought in cognitive psychology that accompanied and all but colonized human factors since its inception during the closing days of the Second World War. Meaning and perceptual order are the end result of an internal trade in representations, representations that get increasingly filled out and meaningful as a result of processing in the mind.

INFORMATION PROCESSING

Information processing did not follow neatly on empiricism, nor did it accompany the initial surge in psychological experimentation. Wundt's introspection did not immediately fuel the development of theoretical substance to fill the gap between elementary matter and the mind's perception of it. Rather, it triggered an anti-subjectivist response that would ban the study of mind and mental processes for decades to come, especially in North America. John Watson, a young psychologist, introduced psychology to the idea of behaviorism in 1913, and aimed to conquer psychology as a purely objective branch of natural science. Introspection was to be disqualified, and any references to or investigations of consciousness were proscribed. The introspective method was seen as unreliable and unscientific, and psychologists had to turn their focus exclusively to phenomena that could be registered and described objectively by independent observers. This meant that introspection had to be replaced by tightly controlled experiments that varied subtle combinations of rewards and punishments in order to bait organisms (anything from mice to pigeons to humans) into particular behaviors over others. The outcome of such experiments was there for all to see, with no need for introspection.

Behaviorism became an early 20th-century embodiment of the Baconian ideal of universal control, this time reflected in a late Industrial Revolution obsession with manipulative technology and domination. It appealed enormously to an optimistic, pragmatic, rapidly developing, and result-oriented North America. Laws extracted from simple experimental settings were thought to carry over to more complex settings and to more experiential phenomena as well, including imagery, thinking, and emotions. Behaviorism was thus fundamentally *nomothetic*: deriving general laws thought to be applicable across people and settings. All human expressions, including art and religion, were reduced to no more than conditioned responses. Behaviorism turned psychology into something wonderfully Newtonian: a schedule of stimuli and responses, of mechanistic, predictable, and changeable couplings between inputs and outputs. The only legitimate characterization of psychology and mental life was one that conformed to the Newtonian framework of classical physics, and abided by its laws of action and reaction.

Then came the Second World War, and the behaviorist bubble was deflated. No matter how clever a system of rewards and punishments psychologists set up, radar operators monitoring for German aircraft intruding into Britain across the Channel would still lose their vigilance over time. They would still have difficulty distinguishing signals from noise, independent of the possible penalties. Pilots would get controls mixed up, and radio operators were evidently limited in their ability to hold information in their heads while getting ready for the next transmission. Where was behaviorism? It could not answer to the new pragmatic appeals. Thus came the first cognitive revolution.

The cognitive revolution reintroduced mind as a legitimate object of study. Rather than manipulating the effect of stimuli on overt responses, it concerned itself with "meaning" as the central concept of psychology. Its aim was, as Bruner (1990) recalled, to discover and describe meanings that people created out of their encounters with the world, and then to propose hypotheses for what meaning-making processes were involved. The very metaphors, however, that legitimized the reintroduction of the study of mind also began to immediately corrupt it. The first cognitive revolution fragmented and became overly technicalized.

The radio and the computer, two technologies accelerated by developments during the Second World War, quickly captured the imagination of those once again studying mental processes. These were formidable similes of mind, able to mechanistically fill the black box (which behaviorism had kept shut) between stimulus and response. The innards of a radio showed filters, channels, and limited capacities through which information flowed. Not much later, all those words appeared in cognitive psychology. Now the mind had filters, channels, and limited capacities too. The computer was even better, containing a working memory, a long-term memory, various forms of storage, input and output, and decision modules. It did not take long for these terms, too, to appear in the psychological lexicon. What seemed to matter most was the ability to quantify and compute mental functioning. Information theory, for example, could explain how elementary stimuli (bits) would flow through processing channels to produce responses. A processed stimulus was deemed informative if it reduced alternative choices, whether the stimulus had to do with Faust or with a digit from a statistical table.

As Bruner (1990) recollected, computability became the necessary and sufficient criterion for cognitive theories. Mind was equated to program. Through these metaphors, the "construction of meaning" quickly became the "processing of information." Newton and Descartes simply would not let go. Once again, psychology was reduced to mechanical components and linkages, and exchanges of energy between and along them. Testing the various components (sensory store, memory, decision making) in endless series of fractionalized laboratory experiments, psychologists hoped, and many are still hoping, that more of the same will eventually add up to something different, that profound insight into the workings of the whole will magically emerge from the study of constituent components.

The Mechanization of Mind

Information processing has been a profoundly Newtonian–Cartesian answer to the mind–matter problem. It is the ultimate mechanization of mind. The basic idea is that the human (mind) is an information processor that takes in stimuli from the outside world, and gradually makes sense of those stimuli by combining them with things already stored in the mind. For example, I see the features of a face, but through coupling them to what I have in long-term memory, I recognize the face as that of my youngest son. Information processing is loyal to the biological-psychological model that sees the matter–mind connection as a physiologically identifiable flow of neuronal energy from periphery to center (from eyeball to cortex), along various nerve pathways. The information-processing pathway of typical models mimics this flow, by taking a stimulus and pushing it through various stages of processing, adding more meaning the entire time. Once the processing system has understood what the stimulus means (or stimuli mean) can an appropriate response be generated (through a backflow from center to periphery, brain to limbs) that in turn creates more (new) stimuli to process.

The Newtonian as well as dualist intimations of the information-processing model are a heartening sight for those with Cartesian anxiety. Thanks to the biological model underneath it, the mind–matter problem is one of a Newtonian transfer (conversion as well as conservation) of energy at all kinds of levels (from photonic energy to nerve impulses, from chemical releases to electrical stimulation, from stimulus to response at the overall organismic level). Both Descartes and Newton can be recognized in the componential explanation of mental functioning (memory, e.g., is typically parsed up in iconic memory, short-term memory, and long-term memory): End-products of mental processing can be exhaustively explained on the basis of interactions between these and other components. Finally of course, information processing is mentalist: It neatly separates *res cogitans* from *res extensa* by studying what happens in the mind entirely separately from what happens in the world. The world is a mere adjunct, truly a *res extensa*, employed solely to lob the next stimulus at the mind (which is where the really interesting processes take place).

The information-processing model works marvelously for the simple laboratory experiments that brought it to life. Laboratory studies of perception, decision making, and reaction time reduce stimuli to single snapshots, fired off at the human processing mechanism as one-stop triggers. The Wundtian idea of awareness as a continually flowing phenomenon is artificially reduced, chunked up and frozen by the very stimuli that subjects are to become aware of. Such dehumanization of the settings in which perception takes place, as well as of the models by which such perception comes about, has given rise to considerable critique. If people are seen to be adding meaning to impoverished, elementary stimuli, then this is because they are given impoverished, elementary, meaningless stimuli in their laboratory tasks! None of that says anything about natural perception or the processes by which people perceive or construct meaning in actual settings. The information-processing model may be true (though even that is judged as unlikely by most), but only for the constrained, Spartan laboratory settings that keep cognition in captivity. If people are seen to struggle in their interpretation of elements, then this may have something to do with the elementary stimuli given to them.

Even Wundt was not without detractors in this respect. The Gestalt movement was launched in part as a response or protest to Wundtian elementarism. Gestaltists claimed that we actually perceive meaningful wholes, that we immediately experience those wholes. We cannot help but see these patterns, these wholes. Max Wertheimer (1880–1934), one of the founding fathers of Gestaltism, illustrates this as such: "I am standing at the window and see a house, trees, sky. And now, for theoretical purposes, I could try to count and say: there are . . . 327 nuances of brightness [and hue]. Do I see '327'? No; I see sky, house, trees" (Wertheimer, cited in

Woods et al., 2002, p. 28). The gestalts that Wertheimer sees (house, trees, sky) are primary to their parts (their elements), and they are more than the sum of their parts. There is an immediate orderliness in experiencing the world. Wertheimer inverts the empiricist claim and information-processing assumption: Rather than meaning being the result of mental operations on elementary stimuli, it actually takes painstaking mental effort (counting 327 nuances of brightness and hue) to reduce primary sensations to their primitive elements. We do not perceive elements: We perceive meaning. Meaning comes effortlessly, prerationally. In contrast, it takes cognitive work to see elements. In the words of William James' senior Harvard colleague Chauncey Wright, there is no antecedent chaos that requires some intrapsychic glue to prevent percepts from falling apart.

New Horizons (Again)

Empiricism does not recognize the immediate orderliness of experience because it does not see relations as real aspects of immediate experience (Heft, 2001). Relations, according to empiricists, are a product of mental (information) processing. This is true for theories of situation awareness. For them, relations between elements are mental artifacts. They get imposed through stages of processing. Subsequent levels of SA add relationships to elements by linking those elements to current meanings and future projections. The problem of the relationship between matter and mind is not at all solved through empiricist responses. But perhaps engineers and designers, as well as many experimental psychologists, are happy to hear about elements (or 327 nuances of brightness and hue), for those can be manipulated in a design prototype and experimentally tested on subjects. Wundt would have done the same thing.

Not unlike Wundt 100 years before him, Ulrich Neisser warned in 1976 that psychology was not quite ready for grand questions about consciousness. Neisser feared that models of cognition would treat consciousness as if it were just a particular stage of processing in a mechanical flow of information. His fears were justified in the mid-1970s, as many psychological models did exactly that. Now they have done it again. Awareness, or consciousness, is equated to a stage of processing along an intrapsychic pathway (levels of SA). As Neisser pointed out, this is an old idea in psychology. The three levels of SA in vogue today were anticipated by Freud, who even supplied flowcharts and boxes in his *Interpretation of Dreams* to map the movements from unconscious (level 1) to preconscious (level 2) to conscious (level 3). The popularity of finding a home, a place, a structure for consciousness in the head is irresistible, said Neisser, as it allows psychology to nail down its most elusive target (consciousness) to a box in a flowchart.

There is a huge cost, though. Along with the deconstruction and mechanization of mental phenomena comes their dehuminization.

Information-processing theories have lost much of their appeal and credibility. Many realize how they have corrupted the spirit of the postbehaviorist cognitive revolution by losing sight of humanity and meaning making. Empiricism (or British empiricism) has slipped into history as a school of thought at the beginning of psychological theorizing. Yet both form legitimate offshoots in current understandings of situation awareness. Notions similar to those of empiricism and information processing are reinvented under new guises, which reintroduces the same type of foundational problems, while leaving some of the really hard problems unaddressed. The problem of the nature of stimuli is one of those, and associated with it is the problem of meaning making. How does the mind make sense of those stimuli? Is meaning the end-product of a processing pathway that flows from periphery to center? These are enormous problems in the history of psychology, all of them problems of the relationship between mind and matter, and all essentially still unresolved. Perhaps they are fundamentally unsolvable within the dualist tradition that psychology inherited from Descartes.

Some movements in human factors are pulling away from the experimental psychological domination. The idea of distributed cognition has renewed the status of the environment as active, constituent participant in cognitive processes, closing the gap between *res cogitans* and *res extensa*. Other people, artifacts, and even body parts are part of the *res cogitans*. How is it otherwise that a child learns to count on his hand, or a soccer player thinks with her feet? Concomitant interest in cognitive work analysis and cognitive systems engineering see such joint cognitive systems as units of analysis, not the constituent human or machine components. Qualitative methods such as ethnography are increasingly legitimate, and critical for understanding distributed cognitive systems. These movements have triggered and embodied what has now become known as the second cognitive revolution, recapturing and rehabilitating the impulses that brought to life the first. How do people make meaning? In order to begin to answer such aboriginal questions, it is now increasingly held as justifiable and necessary to throw the human factors net wider than experimental psychology. Other forms of social inquiry can shed more light on how we are goal-driven creatures in actual, dynamic environments, not passive recipients of snapshot stimuli in a sterile laboratory.

The concerns of these thinkers overlap with functionalist approaches, which formed yet another psychology of protest against Wundt's elementarism. The same protest works equally well against the mechanization of mind by the information-processing school. A century ago, functionalists (William James was one of their foremost exponents) pointed out how people are integrated, living organisms engaged in goal-directed activities, not passive element processors locked into laboratory headrests, buffeted about by one-shot stimuli from an experimental apparatus. The environment in which real activities play out helps shape the organism's responses. Psychological functioning is adaptive: It helps the organism survive and adjust, by incrementally modifying and tweaking its composition or its behavior to generate greater gains on whatever dimension is relevant.

Such ecological thinking is now even beginning to seep into approaches to system safety, which has so far also been dominated by mechanistic, structuralist models (see chap. 2). James was not just a functionalist, however. In fact, he was one of the most all-round psychologists ever. His views on radical empiricism are one great way to access novel thinking about situation awareness and sensemaking, and only appropriate against a background of increasing interest in the role of ecological psychology in human factors.

RADICAL EMPIRICISM

Radical empiricism is one way of circumventing the insurmountable problems associated with psychologies based on dualistic traditions, and William James introduced it as such at the beginning of the 20th century. Radical empiricism rejects the notion of separate mental and material worlds; it rejects dualism. James adhered to an empiricist philosophy, which holds that our knowledge comes (largely) from our discoveries, our experience. But, as Heft (2001) pointed out, James' philosophy is radically empiricist. What is experienced, according to James, is not elements, but relations—meaningful relations. Experienced relations are what perception is made up of. Such a position can account for the orderliness of experience, as it does not rely on subsequent, or *a posteriori* mental processing. Orderliness is an aspect of the ecology, of our world as we experience it and act in it. The world as an ordered, structured universe is experienced, not constructed through mental work. James dealt with the matter–mind problem by letting the knower and the known coincide during the moment of perception (which itself is a constant, uninterrupted flow, rather than a moment). Ontologies (our being in the world) are characterized by continual transactions between knower and known. Order is not imposed on experience, but is itself experienced.

Variations of this approach have always represented a popular countermove in the history of psychology of consciousness. Rather than containing consciousness in a box in the head, it is seen as an aspect of activity. Weick (1995) used the term "enactment" to indicate how people produce the environment they face and are aware of. By acting in the world, people continually create environments that in turn constrain their interpretations, and consequently constrain their next possible actions. This cyclical, ongoing nature of cognition and sensemaking has been recognized by many (see Neisser, 1976) and challenges common interpretations rooted in information-processing psychology where stimuli precede meaning making and (only then) action, and where frozen snapshots of environmental status can be taken as legitimate input to the human processing system. Instead, activities of individuals are only partially triggered by stimuli, because the stimulus itself is produced by activity of the individual.

This moved Weick (1995) to comment that sensemaking never starts; that people are always in the middle of things. Although we may look back on our own experience as consisting of discrete events, the only way to get this impression is to step out of that stream of experience and look down on it from a position of outsider, or retrospective outsider. It is only possible, really, to pay direct attention to what already exists (that which has already passed): "Whatever is now, at the present moment, under way will determine the meaning of whatever has just occurred" (Weick, p. 27). Situation awareness is in part about constructing a plausible story of the process by which an outcome came about, and the reconstruction of immediate history probably plays a dominant role in this. Few theories of situation awareness acknowledge this role, actually, instead directing their analytics to the creation of meaning from elements and the future projection of that meaning.

Radical empiricism does not take the stimulus as its starting point, as does information processing, and neither does it need *a-posteriori* processes (mental, representational) to impose orderliness on sensory impressions. We already experience orderliness and relationships through ongoing, goal-oriented transactions of acting and perceiving. Indeed what we experience during perception is not some cognitive end product in the head. Neisser reminded us of this longstanding issue in 1976 too: Can it be true that we see our own retinal images? The theoretical distance that then needs to be bridged is too large. For if we see that retinal image, who does the looking? Homunculus explanations were unavoidable (and often still are in information processing). Homunculi do not solve the problem of awareness, they simply relocate it.

Rather than a little man in our heads looking at what we are looking at, we ourselves are aware of the world, and its structure, in the world. As Edwin Holt put it, "Consciousness, whenever it is localized at all in space, is not in the skull, but is 'out there' precisely where it appears to be" (cited in Heft, 2001, p. 59). James, and the entire ecological school after him, anticipated this. What is perceived, according to James, is not a replica, not a simile of something out there. What is perceived is already out there. There are no intermediaries between perceiver and percept; perceiving is direct. This position forms the groundwork of ecological approaches in psychology and human factors.

If there is no separation between matter and mind, then there is no gap that needs bridging; there is no need for reconstructive processes in the mind that make sense of elementary stimuli. The Venn diagram with a little and a larger circle that depict actual and ideal situation awareness is superfluous too. Radical empiricism allows human factors to stick closer to the anthropologist's ideal of describing and capturing insider accounts. If there is no separation between mind and matter, between actual and ideal situation awareness, then there is no risk of getting trapped in judging performance by use of extrogenous criteria; criteria imported from outside the setting (informed by hindsight or some other source of omniscience about the situation that opens up that delta, or gap, between what the observer inside the situation knew and what the researcher knows). What the observer inside the situation knows must be seen as canonical— it must be understood not in relation to some normative ideal. For the radical empiricist, there would not be two circles in the Venn diagram, but rather different rationalities, different understandings of the situation— none of them right or wrong or necessarily better or worse, but all of them coupled directly to the interests, expectations, knowledge, and goals of the respective observer.

DRIFTING OFF TRACK: REVISITING A CASE OF "LOST SITUATION AWARENESS"

Let us go back to the Royal Majesty. Traditionalist ideas about a lack of correspondence between a material and a mental world get a boost from this sort of case. A crew ended up 17 miles off track, after a day and a half of sailing. How could this happen? As said before, hindsight makes it easy to see where people were, versus where they thought they were. In hindsight, it is easy to point to the cues and indications that these people should have picked up in order to update or correct or even form their understanding of the unfolding situation around them. Hindsight has a way of exposing those elements that people missed, and a way of amplifying or exaggerating their importance. The key question is not why people did not see what we now know was important. The key question is how they made sense of the situation the way they did. What must the crew in question at the time have seen? How could they, on the basis of their experiences, construct a story that was coherent and plausible? What were the processes by which they became sure that they were right about their position? Let us not question the accuracy of the insider view. Research into situation awareness already does enough of that. Instead, let us try to understand why that insider view was plausible for people at the time, why it was, in fact, the only possible view.

Departure From Bermuda

The Royal Majesty departed Bermuda, bound for Boston at 12:00 noon on the 9th of June 1995. The visibility was good, the winds light, and the sea calm. Before departure the navigator checked the navigation and communication equipment. He found it in "perfect operating condition." About half an hour after departure the harbor pilot disembarked and the course was set toward Boston.

Just before 13:00 there was a cutoff in the signal from the GPS (Global Positioning System) antenna, routed on the fly bridge (the roof of the bridge), to the receiver—leaving the receiver without satellite signals. Postaccident examination showed that the antenna cable had separated from the antenna connection. When it lost satellite reception, the GPS promptly defaulted to dead reckoning (DR) mode. It sounded a brief aural alarm and displayed two codes on its tiny display: DR and SOL. These alarms and codes were not noticed. (DR means that the position is estimated, or deduced, hence "ded," or now "dead," reckoning. SOL means that satellite positions cannot be calculated.) The ship's autopilot would stay in DR mode for the remainder of the journey.

Why was there a DR mode in the GPS in the first place, and why was a default to that mode neither remarkable, nor displayed in a more prominent way on the bridge? When this particular GPS receiver was manufactured (during the 1980s), the GPS satellite system was not as reliable as it is today. The receiver could, when satellite data was unreliable, temporarily use a DR mode in which it estimated positions using an initial position, the gyrocompass for course input and a log for speed input. The GPS thus had two modes, normal and DR. It switched autonomously between the two depending on the accessibility of satellite signals.

By 1995, however, GPS satellite coverage was pretty much complete, and had been working well for years. The crew did not expect anything out of the ordinary. The GPS antenna was moved in February, because parts of the superstructure occasionally would block the incoming signals, which caused temporary and short (a few minutes, according to the captain) periods of DR navigation. This was to a great extent remedied by the antenna move, as the Cruise Line's electronics technician testified. People on the bridge had come to rely on GPS position data and considered other systems to be backup systems. The only times the GPS positions could not be counted on for accuracy were during these brief, normal episodes of signal blockage. Thus, the whole bridge crew was aware of the DR-mode option and how it worked, but none of them ever imagined or were prepared for a sustained loss of satellite data caused by a cable break—no previous loss of satellite data had ever been so swift, so absolute, and so long lasting.

When the GPS switched from normal to DR on this journey in June 1995, an aural alarm sounded and a tiny visual mode annunciation appeared on the display. The aural alarm sounded like that of a digital wristwatch and was less than a second long. The time of the mode change was a busy time (shortly after departure), with multiple tasks and distractors competing for the crew's attention. A departure involves complex maneuvering, there are several crew members on the bridge, and there is a great deal of communication. When a pilot disembarks, the operation is time constrained and risky. In such situations, the aural signal could easily have been drowned out. No one was expecting a reversion to DR mode, and thus the visual indications were not seen either. From the insider perspective, there was no alarm, as there was not going to be a mode default. There was neither a history, nor an expectation of its occurrence.

Yet even if the initial alarm was missed, the mode indication was continuously available on the little GPS display. None of the bridge crew saw it, according to their testimonies. If they had seen it, they knew what it meant, literally translated—dead reckoning means no satellite fixes. But as we saw before, there is a crucial difference between data that in hindsight can be shown to have been available and data that were observable at the time. The indications on the little display (DR and SOL) were placed between two rows of numbers (representing the ship's latitude and longitude) and were about one sixth the size of those numbers. There was no difference in the size and character of the position indications after the switch to DR. The size of the display screen was about 7.5 by 9 centimeters, and the receiver was placed at the aft part of the bridge on a chart table, behind a curtain. The location is reasonable, because it places the GPS, which supplies raw position data, next to the chart, which is normally placed on the chart table. Only in combination with a chart do the GPS data make sense, and furthermore the data were forwarded to the integrated bridge system and displayed there (quite a bit more prominently) as well.

For the crew of the Royal Majesty, this meant that they would have to leave the forward console, actively look at the display, and expect to see more than large digits representing the latitude and longitude. Even then, if they had seen the two-letter code and translated it into the expected behavior of the ship, it is not a certainty that the immediate conclusion would have been "this ship is not heading towards Boston anymore," because temporary DR reversions in the past had never led to such dramatic departures from the planned route. When the officers did leave the forward console to plot a position on the chart, they looked at the display and saw a position, and nothing but a position, because that is what they were expecting to see. It is not a question of them not attending to the indications. They were attending to the indications, the position indications, because plotting the position it is the professional thing to do. For them, the mode change did not exist.

But if the mode change was so nonobservable on the GPS display, why was it not shown more clearly somewhere else? How could one small failure have such an effect—were there no backup systems? The Royal Majesty had a modern integrated bridge system, of which the main component was the navigation and command system (NACOS). The NACOS consisted of two parts, an autopilot part to keep the ship on course and a map construction part, where simple maps could be created and displayed on a radar screen. When the Royal Majesty was being built, the NACOS and the GPS receiver were delivered by different manufacturers, and they, in turn, used different versions of electronic communication standards.

Due to these differing standards and versions, valid position data and invalid DR data sent from the GPS to the NACOS were both labeled with the same code (GP). The installers of the bridge equipment were not told, nor did they expect, that (GP-labeled) position data sent to the NACOS would be anything but valid position data. The designers of the NACOS expected that if invalid data were received, they would have another format. As a result, the GPS used the same data label for valid and invalid data, and thus the autopilot could not distinguish between them. Because the NACOS could not detect that the GPS data was invalid, the ship sailed on an autopilot that was using estimated positions until a few minutes before the grounding.

A principal function of an integrated bridge system is to collect data such as depth, speed, and position from different sensors, which are then shown on a centrally placed display to provide the officer of the watch with an overview of most of the relevant information. The NACOS on the Royal Majesty was placed at the forward part of the bridge, next to the radar screen. Current technological systems commonly have multiple levels of automation with multiple mode indications on many displays. An adaptation of work strategy is to collect these in the same place and another solution is to integrate data from many components into the same display surface. This presents an integration problem for shipping in particular, where quite often components are delivered by different manufacturers.

The centrality of the forward console in an integrated bridge system also sends the implicit message to the officer of the watch that navigation may have taken place at the chart table in times past, but the work is now performed at the console. The chart should still be used, to be sure, but only as a backup option and at regular intervals (customarily every half-hour or every hour). The forward console is perceived to be a clearing house for all the information needed to safely navigate the ship.

As mentioned, the NACOS consisted of two main parts. The GPS sent position data (via the radar) to the NACOS in order to keep the ship on track (autopilot part) and to position the maps on the radar screen (map part). The autopilot part had a number of modes that could be manually selected: NAV and COURSE. NAV mode kept the ship within a certain distance of a track, and corrected for drift caused by wind, sea, and current. COURSE mode was similar but the drift was calculated in an alternative way. The NACOS also had a DR mode, in which the position was continuously estimated. This backup calculation was performed in order to compare the NACOS DR with the position received from the GPS. To calculate the NACOS DR position, data from the gyrocompass and Doppler log were used, but the initial position was regularly updated with GPS data. When the Royal Majesty left Bermuda, the navigation officer chose the NAV mode and the input came from the GPS, normally selected by the crew during the 3 years the vessel had been in service.

If the ship had deviated from her course by more than a preset limit, or if the GPS position had differed from the DR position calculated by the autopilot, the NACOS would have sounded an aural and clearly shown a visual alarm at the forward console (the position-fix alarm). There were no alarms because the two DR positions calculated by the NACOS and the GPS were identical. The NACOS DR, which was the perceived backup, was using GPS data, believed to be valid, to refresh its DR position at regular intervals. This is because the GPS was sending DR data, estimated from log and gyro data, but labeled as valid data. Thus, the radar chart and the autopilot were using the same inaccurate position information and there was no display or warning of the fact that DR positions (from the GPS) were used. Nowhere on the integrated display could the officer on watch confirm what mode the GPS was in, and what effect the mode of the GPS was having on the rest of the automated system, not to mention the ship.

In addition to this, there were no immediate and perceivable effects on the ship because the GPS calculated positions using the log and the gyrocompass. It cannot be expected that a crew should become suspicious of the fact that the ship actually is keeping her speed and course. The combination of a busy departure, an unprecedented event (cable break) together with a nonevent (course keeping), and the change of the locus of navigation (including the intrasystem communication difficulties) shows that it made sense, in the situation and at the time, that the crew did not know that a mode change had occurred.

The Ocean Voyage

Even if the crew did not know about a mode change immediately after departure, there was still a long voyage at sea ahead. Why did none of the officers check the GPS position against another source, such as the Loran-C receiver that was placed close to the GPS? (Loran-C is a radio navigation system that relies on land-based transmitters.) Until the very last minutes before the grounding, the ship did not act strangely and gave no reason for suspecting that anything was amiss. It was a routine trip, the weather was good and the watches and watch changes uneventful.

Several of the officers actually did check the displays of both Loran and GPS receivers, but only used the GPS data (because those had been more reliable in their experience) to plot positions on the paper chart. It was virtually impossible to actually observe the implications of a difference between Loran and GPS numbers alone. Moreover, there were other kinds of cross-checking. Every hour, the position on the radar map was checked against the position on the paper chart, and cues in the world (e.g., sighting of the first buoy) were matched with GPS data. Another subtle reassurance to officers must have been that the master on a number of occasions spent several minutes checking the position and progress of the ship, and did not make any corrections.

Before the GPS antenna was moved, the short spells of signal degradation that lead to DR mode also caused the radar map to jump around on the radar screen (the crew called it "chopping") because the position would change erratically. The reason chopping was not observed on this particular occasion was that the position did not change erratically, but in a manner consistent with dead reckoning. It is entirely possible that the satellite signal was lost before the autopilot was switched on, thus causing no shift in position. The crew had developed a strategy to deal with this occurrence in the past. When the position-fix alarm sounded, they first changed modes (from NAV to COURSE) on the autopilot and then they acknowledged the alarm. This had the effect of stabilizing the map on the radar screen so that it could be used until the GPS signal returned. It was an unreliable strategy, because the map was being used without knowing the extent of error in its positioning on the screen. It also led to the belief that, as mentioned earlier, the only time the GPS data were unreliable was during chopping. Chopping was more or less alleviated by moving the antenna, which means that by eliminating one problem a new pathway for accidents was created. The strategy of using the position-fix alarm as a safeguard no longer covered all or most of the instances of GPS unreliability.

This locally efficient procedure would almost certainly not be found in any manuals, but gained legitimacy through successful repetition becoming common practice over time. It may have sponsored the belief that a stable map is a good map, with the crew concentrating on the visible signs instead of being wary of the errors hidden below the surface. The chopping problem had been resolved for about 4 months, and trust in the automation grew.

First Buoy to Grounding

Looking at the unfolding sequence of events from the position of retrospective outsider, it is once again easy to point to indications missed by the crew. Especially toward the end of the journey, there appears to be a larger number of cues that could potentially have revealed the true nature of the situation. There was an inability of the first officer to positively identify the first buoy that marked the entrance of the Boston sea lanes (such lanes form a separation scheme delineated on the chart to keep meeting and crossing traffic at a safe distance and to keep ships away from dangerous areas). A position error was still not suspected, even with the vessel close to the shore. The lookouts reported red lights and later blue and white water, but the second officer did not take any action. Smaller ships in the area broadcasted warnings on the radio, but nobody on the bridge of the Royal Majesty interpreted those to concern their vessel. The second officer failed to see the second buoy along the sea lanes on the radar, but told the master that it had been sighted. In hindsight, there were numerous opportunities to avoid the grounding, which the crew consistently failed to recognize (NTSB, 1997).

Such conclusions are based on a dualist interpretation of situation awareness. What matters to such an interpretation is the accuracy of the mapping between an external world that can be pieced together in hindsight (and that contains shopping bags full of epiphanies never opened by those who needed them most) and people's internal representation of that world. This internal representation (or situation awareness) can be shown to be clearly deficient, as falling far short of all the cues that were available. But making claims about the awareness of other people at another time and place requires us to put ourselves in their shoes and limit ourselves to what they knew. We have to find out why people thought they were in the right place, or had the right assessment of the situation around them. What made that so? Remember, the adequacy or accuracy of an insider's representation of the situation cannot be called into question: It is what counts for them, and it is what drives further action in that situation. Why was it plausible for the crew to conclude that they were in the right place? What did their world look like to them (not: How does it look now to retrospective observers)?

The first buoy ("BA") in the Boston traffic lanes was passed at 19:20 on the 10th of June, or so the chief officer thought (the buoy identified by the first officer as the BA later turned out to be the "AR" buoy located about 15 miles to the west-southwest of the BA). To the chief officer, there was a buoy on the radar, and it was where he expected it to be, it was where it should be. It made sense to the first officer to identify it as the correct buoy because the echo on the radar screen coincided with the mark on the radar map that signified the BA. Radar map and radar world matched. We now know that the overlap between radar map and radar return was a mere stochastic fit. The map showed the BA buoy, and the radar showed a buoy return. A fascinating coincidence was the sun glare on the ocean surface that made it impossible to visually identify the BA. But independent cross-checking had already occurred: The first officer probably verified his position by two independent means, the radar map and the buoy.

The officer, however, was not alone in managing the situation, or in making sense of it. An interesting aspect of automated navigation systems in real workplaces is that several people typically use it, in partial overlap and consecutively, like the watch-keeping officers on a ship. At 20:00 the second officer took over the watch from the chief officer. The chief officer must have provided the vessel's assumed position, as is good watch-keeping practice. The second officer had no reason to doubt that this was a correct position. The chief officer had been at sea for 21 years, spending 30 of the last 36 months onboard the Royal Majesty. Shortly after the takeover, the second officer reduced the radar scale from 12 to 6 nautical miles. This is normal practice when vessels come closer to shore or other restricted waters. By reducing the scale, there is less clutter from the shore, and an increased likelihood of seeing anomalies and dangers.

When the lookouts later reported lights, the second officer had no expectation that there was anything wrong. To him, the vessel was safely in the traffic lane. Moreover, lookouts are liable to report everything indiscriminately; it is always up to the officer of the watch to decide whether to take action. There is also a cultural and hierarchical gradient between the officer and the lookouts; they come from different nationalities and backgrounds. At this time, the master also visited the bridge and, just after he left, there was a radio call. This escalation of work may well have distracted the second officer from considering the lookouts' report, even if he had wanted to.

After the accident investigation was concluded, it was discovered that two Portuguese fishing vessels had been trying to call the Royal Majesty on the radio to warn her of the imminent danger. The calls were made not long before the grounding, at which time the Royal Majesty was already 16.5 nautical miles from where the crew knew her to be. At 20:42, one of the fishing vessels called, "fishing vessel, fishing vessel call cruise boat," on channel 16 (an international distress channel for emergencies only). Immediately following this first call in English the two fishing vessels started talking to each other in Portuguese. One of the fishing vessels tried to call again a little later, giving the position of the ship he was calling. Calling on the radio without positively identifying the intended receiver can lead to mix-ups. In this case, if the second officer heard the first English call and the ensuing conversation, he most likely disregarded it since it seemed to be two other vessels talking to each other. Such an interpretation makes sense: If one ship calls without identifying the intended receiver, and another ship responds and consequently engages the first caller in conversation, the communication loop is closed. Also, as the officer was using the 6-mile scale, he could not see the fishing vessels on his radar. If he had heard the second call and checked the position, he might well have decided that the call was not for him, as it appeared that he was far from that position. Whomever the fishing ships were calling, it could not have been him, because he was not there.

At about this time, the second buoy should have been seen and around 21:20 it should have been passed, but was not. The second officer assumed that the radar map was correct when it showed that they were on course. To him the buoy signified a position, a distance traveled in the traffic lane, and reporting that it had been passed may have amounted to the same thing as reporting that they had passed the position it was (supposed to have been) in. The second officer did not, at this time, experience an accumulation of anomalies, warning him that something was going wrong. In his view, this buoy, which was perhaps missing or not picked up by the radar, was the first anomaly, but not perceived as a significant one. The typical Bridge Procedures Guide says that a master should be called when (a) something unexpected happens, (b) when something expected does not happen (e.g., a buoy), and (c) at any other time of uncertainty. This is easier to write than it is to apply in practice, particularly in a case where crew members do not see what they expected to see. The NTSB report, in typical counterfactual style, lists at least five actions that the officer should have taken. He did not take any of these actions, because he was not missing opportunities to avoid the grounding. He was navigating the vessel safely to Boston.

The master visited the bridge just before the radio call, telephoned the bridge about 1 hour after it, and made a second visit around 22:00. The times at which he chose to visit the bridge were calm and uneventful, and did not prompt the second officer to voice any concerns, nor did they trigger the master's interest in more closely examining the apparently safe handling of the ship. Five minutes before the grounding, a lookout reported blue and white water. For the second officer, these indications alone were no reason for taking action. They were no warnings of anything about to go amiss, because nothing was going to go amiss. The crew knew where they were. Nothing in their situation suggested to them that they were not doing enough or that they should question the accuracy of their awareness of the situation.

At 22:20 the ship started to veer, which brought the captain to the bridge. The second officer, still certain that they were in the traffic lane, believed that there was something wrong with the steering. This interpretation would be consistent with his experiences of cues and indications during the trip so far. The master, however, came to the bridge and saw the situation differently, but was too late to correct the situation. The Royal Majesty ran aground east of Nantucket at 22:25, at which time she was 17 nautical miles from her planned and presumed course. None of the over 1,000 passengers were injured, but repairs and lost revenues cost the company $7 million.

With a discrepancy of 17 miles at the premature end to the journey of the Royal Majesty, and a day and a half to discover the growing gap between actual and intended track, the case of loss of SA, or deficient SA, looks like it is made. But the supposed elements that make up all the cues and indications that the crew should have seen, and should have understood, are mostly products of hindsight, products of our ability to look at the unfolding sequence of events from the position of retrospective outsiders. In hindsight, we wonder how these repeated "opportunities to avoid the grounding," these repeated invitations to undergo some kind of epiphany about the real nature of the situation, were never experienced by the people who needed them most. But the revelatory nature of the cues, as well as the structure or coherence that they apparently have in retrospect, are not products of the situation itself or the actors in it. They are retrospective imports.

When looked at from the position of retrospective outsider, the deficient SA can look so very real, so compelling. They failed to notice, they did not know, they should have done this or that. But from the point of view of people inside the situation, as well as potential other observers, these deficiencies do not exist in and of themselves; they are artifacts of hindsight, elements removed retrospectively from a stream of action and experience. To people on the inside, it is often nothing more than normal work. If we want to begin to understand why it made sense for people to do what they did, we have to put ourselves in their shoes. What did they know? What was their understanding of the situation? Rather than construing the case as a loss of SA (which simply judges other people for not seeing what we, in our retrospective omniscience, would have seen), there is more explanatory leverage in seeing the crew's actions as normal processes of sensemaking—of transactions between goals, observations, and actions. As Weick (1995) pointed out, sensemaking is

> something that preserves plausibility and coherence, something that is reasonable and memorable, something that embodies past experience and ex-

pectations, something that resonates with other people, something that can be constructed retrospectively but also can be used prospectively, something that captures both feeling and thought . . . In short, what is necessary in sensemaking is a good story. A good story holds disparate elements together long enough to energize and guide action, plausibly enough to allow people to make retrospective sense of whatever happens, and engagingly enough that others will contribute their own inputs in the interest of sensemaking. (p. 61)

Even if one does make concessions to the existence of elements, as Weick does, it is only for the role they play in constructing a plausible story of what is going on, not for building an accurate mental simile of an external world somewhere "out there."

Chapter 6

Why Do Operators Become Complacent?

The introduction of powerful automation in a variety of transport applications has increased the emphasis on human cognitive work. Human operators on, for example, ship bridges or aircraft flight decks spend much time integrating data, planning activities, and managing a suite of machine resources in the conduct of their tasks. This shift has contributed to the utility of a concept such as situation awareness. One large term can capture the extent to which operators are in tune with relevant process data and can form a picture of the system and its progress in space and time.

As the Royal Majesty example in the chapter 5 showed, most high-tech settings are actually not characterized by a single human interacting with a machine. In almost all cases, multiple people—crews, or teams of operators—jointly interact with the automated system in pursuit of operational objectives. These crews or teams have to coordinate their activities with those of the system in order to achieve common goals.

Despite the weight that crews (and human factors researchers) repeatedly attribute to having a shared understanding of their system state and problems to be solved, consensus in transportation human factors on a concept of crew situation awareness seems far off. It appears that various labels are used interchangeably to refer to the same basic phenomenon, for example, group situation awareness, shared problem models, team situation awareness, mutual knowledge, shared mental models, joint situation awareness, and shared understanding. At the same time, results about what constitutes the phenomenon are fragmented and ideas on how to measure it remain divided. Methods to gain empirical access range from modified measures of practitioner expertise, to questionnaires interjected into suddenly frozen simulation scenarios, to implicit probes embedded in unfolding simulations of natural task behavior.

Most critically, however, a common definition or model of crew situation awareness remains elusive. There is human factors research, for example, that claims to identify links between crew situation awareness and other parameters (such as planning or crew-member roles). But such research often does not mention a definition of the phenomenon. This renders empirical demonstrations of the phenomenon unverifiable and inconclusive. After all, how can a researcher claim that he or she saw something if that something was not defined? Perhaps there is no need to define the phenomenon, because everybody knows what it means. Indeed, situation awareness is what we call a *folk model*. It has come up from the practitioner community (fighter pilots in this case) to indicate the degree of coupling between human and environment. Folk models are highly useful because they can collapse complex, multidimensional problems into simple labels that everybody can relate to. But this is also where the risks lie, certainly when researchers pick up on a folk label and attempt to investigate and model it scientifically.

Situation awareness is not alone in this. Human factors today has more concepts that aim to provide insight into the human performance issues that underlie complex behavioral sequences. It is often tempting to mistake the labels themselves for deeper insight—something that is becoming increasingly common in, for example, accident analyses. Thus loss of situation awareness, automation complacency and loss of effective crew resource management can now be found among the causal factors and conclusions in accident reports. This happens without further specification of the psychological mechanisms responsible for the observed behavior—much less how such mechanisms or behavior could have forced the sequence of events toward its eventual outcome. The labels (modernist replacements of the old *pilot error*) are used to refer to concepts that are intuitively meaningful. Everyone is assumed to understand or implicitly agree on them, yet no effort is usually made to explicate or reach agreement on the underlying mechanisms or precise definitions. People may no longer dare to ask what these labels mean, lest others suspect they are not really initiated in the particulars of their business.

Indeed, large labels that correspond roughly to mental phenomena we know from daily life are deemed sufficient—they need no further explanation. This is often accepted practice for psychological phenomena because as humans we all have privileged knowledge about how the mind works (because we all have one). However, a verifiable and detailed mapping between the context-specific (and measurable) particulars of a behavior on the one hand and a concept-dependent model on the other is not achieved—the jump from context specifics (somebody flying into a mountainside) to concept dependence (the operator must have lost SA) is immune to critique or verification.

Folk models are not necessarily incorrect, but compared to articulated models they focus on descriptions rather than explanations, and they are very hard to prove wrong. Folk models are pervasive in the history of science. One well-known example of a folk model from modern times is Freud's psychodynamic model, which links observable behavior and emotions to nonobservable structures (id, ego, superego) and their interactions. One feature of folk models is that nonobservable constructs are endowed with the necessary causal power without much specification of the mechanism responsible for such causation. According to Kern (1998), for example, complacency can cause a loss of situation awareness. In other words, one folk problem causes another folk problem. Such assertions leave few people any wiser. Because both folk problems are constructs postulated by outside observers (and mostly post hoc), they cannot logically cause anything in the empirical world. Yet this is precisely what they are assumed to be capable of. In wrapping up a conference on situation awareness, Charles Billings warned against this danger in 1996:

> The most serious shortcoming of the situation awareness construct as we have thought about it to date, however, is that it's too neat, too holistic and too seductive. We heard here that deficient SA was a causal factor in many airline accidents associated with human error. We must avoid this trap: deficient situation awareness doesn't "cause" anything. Faulty spatial perception, diverted attention, inability to acquire data in the time available, deficient decision-making, perhaps, but not a deficient abstraction! (p. 3)

What Billings did not mention is that "diverted attention" and "deficient decision-making" themselves are abstractions at some level (and post hoc ones at that). They are nevertheless less contentious because they provide a reasonable level of detail in their description of the psychological mechanisms that account for their causation. Situation awareness is too "neat" and "holistic" in the sense that it lacks such a level of detail and thus fails to account for a psychological mechanism that would connect features of the sequence of events to the outcome. The folk model, however, was coined precisely because practitioners (pilots) wanted something "neat" and "holistic" that could capture critical but inexplicit aspects of their performance in complex, dynamic situations. We have to see their use of a folk model as legitimate. It can fulfill a useful function with respect to the concerns and goals of a user community.

This does not mean that the concepts coined by users can be taken up and causally manipulated by scientists without serious foundational analysis and explication of their meaning. Resisting the temptation, however, can be difficult. After all, human factors is a discipline that lives by its applied usefulness. If the discipline does not generate anything of interest to applied communities, then why would they bother funding the work? In this sense, folk models can seem like a wonderfully convenient bridge between basic and applied worlds, between scientific and practitioner communities. Terms like *situation awareness* allow both camps to speak the same language. But such conceptual sharing risks selling out to superficial validity. It may not do human factors a lot of good in the long run, nor may it really benefit the practitioner consumers of research results.

Another folk concept is complacency. Why does people's vigilance decline over time, especially when confronted with repetitive stimuli? Vigilance decrements have formed an interesting research problem ever since the birth of human factors during and just after the Second World War. The idea of complacency has always been related to vigilance problems. Although complacency connotes something motivational (people must ensure that they watch the process carefully), the human factors literature actually has little in the way of explanation or definition. What is complacency? Why does it occur? If you want answers to these questions, do not turn to the human factors literature. You will not find answers there. Complacency is one of those constructs, whose meaning is assumed to be known by everyone. This justifies taking it up in scientific discourse as something that can be manipulated or studied as an independent or dependent variable, without having to go through the bother of defining what it actually is or how it works. In other words, complacency makes a "neat" and "holistic" case for studying folk models.

DEFINITION BY SUBSTITUTION

The most evident characteristic of folk models is that they define their central constructs by substitution rather than decomposition. A folk concept is explained simply by referring to another phenomenon or construct that itself is in equal need of explanation. Substitution is not the same as decomposition: Substituting replaces one high-level label with another, whereas decomposition takes the analysis down into subsequent levels of greater detail, which transform the high-level concept into increasingly measurable context specifics. A good example of definition by substitution is the label complacency, in relation to the problems observed on automated flight decks. Most textbooks on aviation human factors talk about complacency and even endow it with causal power, but none really define (i.e., decompose) it:

- According to Wiener (1988, p. 452), "boredom and complacency are often mentioned" in connection with the out-of-the-loop issue in automated cockpits. But whether complacency causes an out-of-the-loop condition or whether it is the other way around is left unanswered.

• O'Hare and Roscoe (1990, p. 117) stated that "because autopilots have proved extremely reliable, pilots tend to become complacent and fail to monitor them." Complacency, in other words, is invoked to explain monitor failures.

• Kern (1998, p. 240) maintained that "as pilots perform duties as system monitors they will be lulled into complacency, lose situational awareness, and not be prepared to react in a timely manner when the system fails." Thus, complacency can cause a loss of situational awareness. But how this occurs is left to the imagination.

• On the same page in their textbook, Campbell and Bagshaw (1991, p. 126) said that complacency is both a "*trait* that can lead to a reduced awareness of danger," and a "*state* of confidence plus contentment" (emphasis added). In other words, complacency is at the same time a long-lasting, enduring feature of personality (a trait) and a shorter lived, transient phase in performance (a state).

• For the purpose of categorizing incident reports, Parasuraman, Molly, and Singh (1993, p. 3) defined complacency as: "self-satisfaction which may result in non-vigilance based on an unjustified assumption of satisfactory system state." This is part definition but also part substitution: Self-satisfaction takes the place of complacency and is assumed to speak for itself. There is no need to make explicit by which psychological mechanism self-satisfaction arises or how it produces nonvigilance.

It is in fact difficult to find real content on complacency in the human factors literature. The phenomenon is often described or mentioned in relation to some deviation or diversion from official guidance (people should coordinate, double-check, look—but they do not), which is both normativist and judgmental. The "unjustified assumption of satisfactory system state" in Parasuraman et al.'s (1993) definition is emblematic for human factors' understanding of work by reference to externally dictated norms. If we want to understand complacency, the whole point is to analyze why the assumption of satisfactory system state is justified (not unjustified) by those who are making that assumption. If it were unjustified, and they knew that, they would not make the assumption and would consequently not become complacent. Saying that an assumption of satisfactory system state is unjustified (but people still keep making it—they must be motivationally deficient) does not explain much at all.

None of the above examples really provide a definition of complacency. Instead, complacency is treated as self-evident (everybody knows what it means, right?) and thus it can be defined by substituting one label for another. The human factors literature equates complacency with many different labels, including boredom, overconfidence, contentment, unwarranted faith, overreliance, self-satisfaction, and even a low index of suspicion. So if we would ask, "What do you mean by 'complacency'?," and the reply is, "Well, it is self-satisfaction," we can be expected to say, "Oh, of course, now I understand what you mean." But do we really? Explanation by substitution actually raises more questions than it answers. By failing to propose an articulated psychological mechanism responsible for the behavior observed, we are left to wonder. How is it that complacency produces vigilance decrements or how is it that complacency leads to a loss of situation awareness? The explanation could be a decay of neurological connections, fluctuations in learning and motivation, or a conscious trade-off between competing goals in a changing environment. Such definitions, which begin to operationalize the large concept of complacency, suggest possible probes that a researcher could use to monitor for the target effect. But because none of the descriptions of complacency available today offer any such roads to insight, claims that complacency was at the heart of a sequence of events are immune to critique and falsification.

IMMUNITY AGAINST FALSIFICATION

Most philosophies of science rely on the empirical world as touchstone or ultimate arbiter (a reality check) for postulated theories. Following Popper's rejection of the inductive method in the empirical sciences, theories and hypotheses can only be deductively validated by means of falsifiability. This usually involves some form of empirical testing to look for exceptions to the postulated hypothesis, where the absence of contradictory evidence becomes corroboration of the theory. Falsification deals with the central weakness of the inductive method of verification, which, as pointed out by David Hume, requires an infinite number of confirming empirical demonstrations. Falsification, on the other hand, can work on the basis of only one empirical instance, which proves the theory wrong. As seen in chapter 3, this is of course a highly idealized, almost clinical conceptualization of the scientific enterprise. Yet, regardless, theories that do not permit falsification at all are highly suspect.

The resistance of folk models against falsification is known as *immunization*. Folk models leave assertions about empirical reality underspecified, without a trace for others to follow or critique. For example, a senior training captain once asserted that cockpit discipline is compromised when any of the following attitudes are prevalent: arrogance, complacency, and overconfidence. Nobody can disagree because the assertion is underspecified and therefore immune against falsification. This is similar to psychoanalysts claiming that obsessive-compulsive disorders are the result of overly harsh toilet training that fixated the individual in the anal stage. In the same vein, if the question of "Where are we headed?" from one pilot to the other is interpreted as a loss of situation awareness (Aeronautica Civil, 1996), this claim is immune against falsification. The journey from context-specific behavior (people asking questions) to the postulated psychological mechanism (loss of situation awareness) is made in one big leap, leaving no trace for others to follow or critique.

Current theories of situation awareness are not sufficiently articulated to be able to explain why asking questions about direction represents a loss of situation awareness. Some theories may superficially appear to have the characteristics of good scientific models, yet just below the surface they lack an articulated mechanism that is amenable to falsification. Although falsifiability may at first seem like a self-defeating criterion for scientific progress, the opposite is true: The most falsifiable models are usually also the most informative ones, in the sense that they make stronger and more demonstrable claims about reality. In other words, falsifiability and informativeness are two sides of the same coin.

Folk Models Versus Young and Promising Models

One risk in rejecting folk models is that the baby is thrown out with the bath water. In other words, there is the risk of rejecting even those models that may be able to generate useful empirical results, if only given the time and opportunity to do so. Indeed, the more articulated human factors constructs (such as decision making, diagnosis) are distinguished from the less articulated ones (situation awareness, complacency) in part by their maturity, by how long they have been around in the discipline. What opportunity should the younger ones receive before being rejected as unproductive? The answer to this question hinges, once again, on falsifiability. Ideal progress in science is described as the succession of theories, each of which is more falsifiable (and thus more informative) than the one before it. Yet when we assess loss of situation awareness or complacency as more novel explanations of phenomena that were previously covered by other explanations, it is easy to see that falsifiability has actually decreased, rather than increased.

Take as an example an automation-related accident that occurred when situation awareness or automation-induced complacency did not yet exist—in 1973. The aircraft in question was on approach in rapidly changing weather conditions. It was equipped with a slightly deficient *flight director* (a device on the central instrument panel showing the pilot where to go, based on an unseen variety of sensory inputs), which the captain of the airplane distrusted. The airplane struck a seawall bounding Boston's Logan Airport about 1 kilometer short of the runway and slightly to the side of it, killing all 89 people onboard. In its comment on the crash, the National Transportation Safety Board explained how an accumulation of discrepancies, none critical in themselves, can rapidly deteriorate into a high-risk situation without positive flight management. The first officer, who was flying, was preoccupied with the information presented by his flight-director systems, to the detriment of his attention to altitude, heading and airspeed control (NTSB, 1974).

Today, both automation-induced complacency of the first officer and a loss of situation awareness of the entire crew could likely be cited under the causes of this crash. (Actually, that the same set of empirical phenomena can comfortably be grouped under either label—complacency or loss of situation awareness—is additional testimony to the undifferentiated and underspecified nature of these concepts.) These supposed explanations (complacency, loss of situation awareness) were obviously not needed in 1974 to deal with this accident. The analysis left us instead with more detailed, more falsifiable, and more traceable assertions that linked features of the situation (e.g., an accumulation of discrepancies) with measurable or demonstrable aspects of human performance (diversion of attention to the flight director vs. other sources of data). The decrease of falsifiability represented by complacency and situation awareness as hypothetical contenders in explaining this crash represents the inverse of scientific progress, and therefore argues for the rejection of such novel concepts.

OVERGENERALIZATION

The lack of specificity of folk models and the inability to falsify them contribute to their overgeneralization. One famous example of overgeneralization in psychology is the inverted-U curve, also known as the Yerkes–Dodson law. Ubiquitous in human factors textbooks, the inverted-U curve couples arousal with performance (without clearly stating any units of either arousal or performance), where a person's best performance is claimed to occur between too much arousal (or stress) and too little, tracing a sort of hyperbole. The original experiments were, however, neither about performance nor about arousal (Yerkes & Dodson, 1908). They were not even about humans. Examining "the relation between stimulus strength and habit formation," the researchers subjected laboratory rats to electrical shocks to see how quickly they decided to take a particular pathway versus another. The conclusion was that rats learn best (that is, they form habits most rapidly) at any but the highest or lowest shock. The results approximated an inverted U only with a most generous curve fitting, the x axis was never defined in psychological terms but in terms of shock strength, and even this was confounded: Yerkes and Dodson used different levels of shock which were too poorly calibrated to know how different they really were. The subsequent overgeneralization of the Yerkes–Dodson results (to no fault of their own, incidentally) has confounded stress and arousal, and after a century there is still little evidence that any kind of inverted-U relationship holds for stress (or arousal) and human performance. Overgeneralizations take narrow laboratory findings and apply them uncritically to any broad situation where behavioral particulars bear some prima-facie resemblance to the phenomenon that was investigated under controlled circumstances.

Other examples of overgeneralization and overapplication include perceptual tunneling (putatively championed by the crew of an airliner that descended into the Everglades after its autopilot was inadvertently switched off) and the loss of effective Crew Resource Management (CRM) as major explanations of accidents (e.g., Aeronautica Civil, 1996). A most frequently quoted sequence of events with respect to CRM is the flight of an iced-up airliner from Washington National Airport in the winter of 1982 that ended shortly after takeoff on the 14th Street bridge and in the Potomac River. The basic cause of the accident is said to be the copilot's unassertive remarks about an irregular engine instrument reading (despite the fact that the copilot was known for his assertiveness). This supposed explanation hides many other factors which might be more relevant, including air-traffic control pressures, the controversy surrounding rejected takeoffs close to decision speed, the sensitivity of the aircraft type to icing and its pitch-up tendency with even little ice on the slats (devices on the wing's leading edge that help it fly at slow speeds), and ambiguous engineering language in the airplane manual to describe the conditions for use of engine anti-ice.

In an effort to explain complex behavior, and still make a connection to the applied worlds from which it owes its existence, transportation human factors may be doing itself a disservice by inventing and uncritically using folk models. If we use models that do not articulate the performance measures that can be used in the particular contexts that we want to speak about, we can make no progress in better understanding the sources of success and failure in our operational environments.

Chapter 7

Why Don't They Follow
the Procedures?

People do not always follow procedures. We can easily observe this when watching people at work, and managers, supervisors and regulators (or anybody else responsible for safe outcomes of work) often consider it to be a large practical problem. In chapter 6 we saw how *complacency* would be a very unsatisfactory label for explaining practical drift away from written guidance. But what lies behind it then?

In hindsight, after a mishap, rule violations seem to play such a dominant causal role. If only they had followed the procedure! Studies keep returning the basic finding that procedure violations precede accidents. For example, an analysis carried out for an aircraft manufacturer identified "pilot deviation from basic operational procedure" as primary factor in almost 100 accidents (Lautman & Gallimore, 1987, p. 2). One methodological problem with such work is that it selects its cases on the dependent variable (the accident), thereby generating tautologies rather than findings. But performance variations, especially those at odds with written guidance, easily get overestimated for their role in the sequence of events:

> The interpretation of what happened may then be distorted by naturalistic biases to overestimate the possible causal role of unofficial action or procedural violation. . . . While it is possible to show that violations of procedures are involved in many safety events, many violations of procedures are not, and indeed some violations (strictly interpreted) appear to represent more effective ways of working. (McDonald, Corrigan, & Ward, 2002, pp. 3–5)

As seen in chapter 4, hindsight turns complex, tangled histories laced with uncertainty and pressure into neat, linear anecdotes with obvious choices. What look like violations from the outside and hindsight are often actions that make sense given the pressures and trade-offs that exist on this inside of real work. Finding procedure violations as causes or contributors to mishaps, in other words, says more about us, and the biases we introduce when looking back on a sequence of events, than it does about people who were doing actual work at the time.

Yet if procedure violations are judged to be such a large ingredient of mishaps, then it can be tempting, in the wake of failure, to introduce even more procedures, or to change existing ones, or to enforce stricter compliance. For example, shortly after a fatal shootdown of two U.S. Black Hawk helicopters over Northern Iraq by U.S. fighter jets, "higher headquarters in Europe dispatched a sweeping set of rules in documents several inches thick to 'absolutely guarantee' that whatever caused this tragedy would never happen again" (Snook, 2000, p. 201). It is a common, but not typically satisfactory, reaction. Introducing more procedures does not necessarily avoid the next incident, nor do exhortations to follow rules more carefully necessarily increase compliance or enhance safety. In the end, a mismatch between procedures and practice is not unique to accident sequences. Not following procedures does not necessarily lead to trouble, and safe outcomes may be preceded by just as many procedural deviations as accidents are.

PROCEDURE APPLICATION AS RULE-FOLLOWING

When rules are violated, are these bad people ignoring the rules? Or are these bad rules, ill matched to the demands of real work? To be sure, procedures, with the aim of standardization, can play an important role in shaping safe practice. Commercial aviation is often held up as a prime example of the powerful effect of standardization on safety. But there is a deeper, more complex dynamic where real practice is continually adrift from official written guidance, settling at times, unsettled and shifting at others. There is a deeper, more complex interplay whereby practice sometimes precedes and defines the rules rather than being defined by them. In those cases, is a violation an expression of defiance, or an expression of compliance—people following practical rules rather than official, impractical ones?

These possibilities lie between two opposing models of what procedures mean, and what they in turn mean for safety. These models of procedures guide how organizations think about making progress on safety. The first model is based on the notion that not following procedures can lead to unsafe situations. These are its premises:

- Procedures represent the best thought-out, and thus the safest, way to carry out a job.
- Procedure following is mostly simple IF–THEN rule-based mental activity: IF this situation occurs, THEN this algorithm (e.g., checklist) applies.
- Safety results from people following procedures.
- For progress on safety, organizations must invest in people's knowledge of procedures and ensure that procedures are followed.

In this idea of procedures, those who violate them are often depicted as putting themselves above the law. These people may think that rules procedures are made for others, but not for them, as they know how to really do the job. This idea of rules and procedures suggests that there is something exceptionalist or misguidedly elitist about those who choose not to follow the rules. After a maintenance related mishap, for example, investigators found that "the engineers who carried out the flap change demonstrated a willingness to work around difficulties without reference to the design authority, including situations where compliance with the maintenance manual could not be achieved" (Joint Aviation Authorities, 2001). The engineers demonstrated a "willingness." Such terminology embodies notions of volition (the engineers had a free choice either to comply or not) and full rationality (they knew what they were doing). They violated willingly. Violators are wrong, because rules and procedures prescribe the best, safest way to do a job, independent of who does that job. Rules and procedures are for everyone.

Such characterizations are naive at best, and always misleading. If you know where to look, daily practice is testimony to the ambiguity of procedures, and evidence that procedures are a rather problematic category of human work. First, real work takes place in a context of limited resources and multiple goals and pressures. Procedures assume that there is time to do them in, certainty (of what the situation is), and sufficient information available (e.g., about whether tasks are accomplished according to the procedure). This already keeps rules at a distance from actual tasks, because real work seldom meets those criteria. Work-to-rule strikes show how it can be impossible to follow the rules and get the job done at the same time.

Aviation line maintenance is emblematic: A job-perception gap exists where supervisors are convinced that safety and success result from mechanics following procedures—a sign-off means that applicable procedures were followed. But mechanics may encounter problems for which the right tools or parts are not at hand; the aircraft may be parked far away from base. Or there may be too little time: Aircraft with a considerable number of problems may have to be turned around for the next flight within half an hour. Mechanics, consequently, see success as the result of their evolved skills at adapting, inventing, compromising, and improvising in the face of local pressures and challenges on the line—a sign-off means the job was accomplished in spite of resource limitations, organizational dilemmas, and pressures. Those mechanics who are most adept are valued for their productive capacity even by higher organizational levels. Unacknowledged by those levels, though, are the vast informal work systems that develop so mechanics can get work done, advance their skills at improvising and satisficing, impart them to one another, and condense them in unofficial, self-made documentation (McDonald et al., 2002). Seen from the outside, a defining characteristic of such informal work systems would be routine nonconformity. But from the inside, the same behavior is a mark of expertise, fueled by professional and interpeer pride. And of course, informal work systems emerge and thrive in the first place because procedures are inadequate to cope with local challenges and surprises, and because procedures' conception of work collides with the scarcity, pressure and multiple goals of real work.

Some of the safest complex, dynamic work not only occurs despite the procedures—such as aircraft line maintenance—but without procedures altogether. Rochlin et al. (1987, p. 79), commenting on the introduction of ever heavier and capable aircraft onto naval aircraft carriers, noted that "there were no books on the integration of this new hardware into existing routines and no other place to practice it but at sea. Moreover, little of the process was written down, so that the ship in operation is the only reliable manual." Work is "neither standardized across ships nor, in fact, written down systematically and formally anywhere." Yet naval aircraft carriers, with inherent high-risk operations, have a remarkable safety record, like other so-called high-reliability organizations (Rochlin, 1999; Rochlin, LaPorte, & Roberts, 1987). Documentation cannot present any close relationship to situated action because of the unlimited uncertainty and ambiguity involved in the activity. Especially where normal work mirrors the uncertainty and criticality of emergencies, rules emerge from practice and experience rather than preceding it. Procedures, in other words, end up following work instead of specifying action beforehand. Human factors has so far been unable to trace and model such coevolution of human and system, of work and rules. Instead, it has typically imposed a mechanistic, static view of one best practice from the top down.

Procedure-following can also be antithetical to safety. In the 1949 U.S. Mann Gulch disaster, firefighters who perished were the ones sticking to the organizational mandate to carry their tools everywhere (Weick, 1993). In this case, as in others (e.g., Carley, 1999), people faced the choice between following the procedure or surviving.

Procedures Are Limited in Rationalizing Human Work

This, then, is the tension. Procedures are seen as an investment in safety—but it turns out that they not always are. Procedures are thought to be required to achieve safe practice—yet they are not always necessary, nor likely ever sufficient for creating safety. Procedures spell out how to do the job safely—yet following all the procedures can lead to an inability to get the job done. Though a considerable practical problem, such tensions are underreported and underanalyzed in the human factors literature.

There is always a distance between a written rule and an actual task. This distance needs to be bridged; the gap must be closed, and the only thing that can close it is human interpretation and application. Ethnographer Ed Hutchins has pointed out how procedures are not just externalized cognitive tasks (Wright & McCarthy, 2003). Externalizing a cognitive task would transplanted it from the head to the world, for example onto a checklist. Rather, following a procedure requires cognitive tasks that are not specified in the procedure; transforming the written procedure into activity requires cognitive work. Procedures are inevitably incomplete specifications of action: They contain abstract descriptions of objects and actions that relate only loosely to particular objects and actions that are encountered in the actual situation (Suchman, 1987). Take as an example the lubrication of the jackscrew on MD-80s from chapter 2—something that was done incompletely and at increasingly greater intervals before the crash of Alaska 261. This is part of the written procedure that describes how the lubrication work should be done (NTSB, 2002, pp. 29–30):

A. Open access doors 6307, 6308, 6306 and 6309
B. Lube per the following . . .
    3. JACKSCREW
    Apply light coat of grease to threads, then operate mechanism through full range of travel to distribute lubricant over length of jackscrew.
C. Close doors 6307, 6308, 6306 and 6309

This leaves a lot to the imagination, or to the mechanic's initiative. How much is a "light" coat? Do you do apply the grease with a brush (if a "light coat" is what you need), or do you pump it onto the parts directly with the grease gun? How often should the mechanism (jackscrew plus nut) be operated through its full range of travel during the lubrication procedure? None of this is specified in the written guidance. It is little wonder that:

> Investigators observed that different methods were used by maintenance personnel to accomplish certain steps in the lubrication procedure, including the manner in which grease was applied to the acme nut fitting and the acme screw and the number of times the trim system was cycled to distribute the grease immediately after its application. (NTSB, 2002, p. 116)

In addition, actually carrying out the work is difficult enough. As noted in chapter 2, the access panels of the horizontal stabilizer were just large enough to allow a hand through, which would then block the view of anything that went on inside. As a mechanic, you can either look at what you have to do or what you have just done, or actually do it. You cannot do both at the same time, because the access doors are too small. This makes judgments about how well the work is being done rather difficult. The investigation discovered as much when they interviewed the mechanic responsible for the last lubrication of the accident airplane: "When asked how he determined whether the lubrication was being accomplished properly and when to stop pumping the grease gun, the mechanic responded, 'I don't' " (NTSB, 2002, p. 31).

The time the lubrication procedure took was also unclear, as there was ambiguity about which steps were included in the procedure. Where does the procedure begin and where does it end, after access has been created to the area, or before? And is closing the panels part of it as well, as far as time estimates are concerned? Having heard that the entire lubrication process takes "a couple of hours," investigators learned from the mechanic of the accident airplane that:

> the lubrication task took "roughly . . . probably an hour" to accomplish. It was not entirely clear from his testimony whether he was including removal of the access panels in his estimate. When asked whether his 1-hour estimate included gaining access to the area, he replied, "No, that would probably take a little—well, you've got probably a dozen screws to take out of the one panel, so that's—I wouldn't think any more than an hour." The questioner then stated, "including access?," and the mechanic responded, "Yeah." (NTSB, 2002, p. 32)

As the procedure for lubricating the MD-80 jackscrew indicates, and McDonald et al. (2002) remind us, formal documentation cannot be relied on, nor is it normally available in a way which supports a close relationship to action. There is a distinction between universalistic and particularistic rules: Universalistic rules are very general prescriptions (e.g., "Apply light coat of grease to threads"), but remain at a distance from their actual application. In fact, all universalistic rules or general prescriptions develop into particularistic rules as experience accumulates. With experience, people encounter the conditions under which universalistic rules need to be applied, and become increasingly able to specify those conditions. As a result, universalistic rules assume appropriate local expressions through practice.

Wright and McCarthy (2003) have pointed out that procedures come out of the scientific management tradition, where their main purpose was a minimization of human variability, maximization of predictability, a rationalization of work. Aviation contains a strong heritage: Procedures in commercial aviation represent and allow a routinization that makes it possible to conduct safety-critical work with perfect strangers. Procedures are a substitute for knowing coworkers. The actions of a copilot are predictable not because the copilot is known (in fact, you may never have flown with him or her), but because the procedures make them predictable. Without such standardization it would be impossible to cooperate safely and smoothly with unknown people.

In the spirit of scientific management, human factors also assumes that order and stability in operational systems are achieved rationally, mechanistically, and that control is implemented vertically (e.g., through task analyses that produce prescriptions of work to be carried out). In addition, the strong influence of information-processing psychology on human factors has reinforced the idea of procedures as IF–THEN rule following, where procedures are akin to a program in a computer that in turn serves as input signals to the human information processor. The algorithm specified by the procedure becomes the software on which the human processor runs. But it is not that simple. Following procedures in the sense of applying them in practice requires more intelligence. It requires additional cognitive work. This brings us to the second model of procedures and safety.

PROCEDURE APPLICATION AS SUBSTANTIVE COGNITIVE ACTIVITY

People at work must interpret procedures with respect to a collection of actions and circumstances that the procedures themselves can never fully specify (e.g., Suchman, 1987). In other words, procedures are not the work itself. Work, especially that in complex, dynamic workplaces, often requires subtle, local judgments with regard to timing of subtasks, relevance, importance, prioritization, and so forth. For example, there is no technical reason why a before-landing checklist in a commercial aircraft could not be automated. The kinds of items on such a checklist (e.g., hydraulic pumps, gear, flaps) are mostly mechanical and could be activated on the basis of predetermined logic without having to rely on, or constantly remind, a human to do so. Yet no before-landing checklist is fully automated today. The reason is that approaches for landing differ—they can differ in terms of timing, workload, or other priorities. Indeed, the reason is that the checklist is not the job itself. The checklist is, to repeat Suchman, a resource for action; it is one way for people to help structure activities across roughly similar yet subtly different situations. Variability in this is inevitable. Circumstances change, or are not as foreseen by those who designed the procedures. Safety, then, is not the result of rote rule following; it is the result of people's insight into the features of situations that demand certain actions, and people being skillful at finding and using a variety of resources (including written guidance) to accomplish their goals. This suggests a second model on procedures and safety:

- Procedures are resources for action. Procedures do not specify all circumstances to which they apply. Procedures cannot dictate their own application.
- Applying procedures successfully across situations can be a substantive and skillful cognitive activity.
- Procedures cannot, in themselves, guarantee safety. Safety results from people being skillful at judging when and how (and when not) to adapt procedures to local circumstances.
- For progress on safety, organizations must monitor and understand the reasons behind the gap between procedures and practice. Additionally, organizations must develop ways that support people's skill at judging when and how to adapt.

Procedures and Unusual Situations

Although there is always a distance between the logics dictated in written guidance and real actions to be taken in the world, prespecified guidance is especially inadequate in the face of novelty and uncertainty. Adapting procedures to fit unusual circumstances is a substantive cognitive activity. Take for instance the crash of a large passenger aircraft near Halifax, Nova Scotia in 1998. After an uneventful departure, a burning smell was detected and, not much later, smoke was reported inside the cockpit. Carley (1999) characterized the two pilots as respective embodiments of the models of procedures and safety: The co-pilot preferred a rapid descent and suggested dumping fuel early so that the aircraft would not be too heavy to land. But the captain told the copilot, who was flying the plane, not to descend too fast, and insisted they cover applicable procedures (checklists) for dealing with smoke and fire. The captain delayed a decision on dumping fuel. With the fire developing, the aircraft became uncontrollable and crashed into the sea, taking all 229 lives onboard with it. There were many good reasons for not immediately diverting to Halifax: Neither pilot was familiar with the airport, they would have to fly an approach procedure that they were not very proficient at, applicable charts and information on the airport were not easily available, and an extensive meal service had just been started in the cabin.

Yet, part of the example illustrates a fundamental double bind for those who encounter surprise and have to apply procedures in practice (Woods & Shattuck, 2000):

- If rote rule following persists in the face of cues that suggest procedures should be adapted, this may lead to unsafe outcomes. People can get blamed for their inflexibility, their application of rules without sensitivity to context.
- If adaptations to unanticipated conditions are attempted without complete knowledge of circumstance or certainty of outcome, unsafe results may occur too. In this case, people get blamed for their deviations their nonadherence.

In other words, people can fail to adapt, or attempt adaptations that may fail. Rule following can become a desynchronized and increasingly irrelevant activity, decoupled from how events and breakdowns are really unfolding and multiplying throughout a system. In the Halifax crash, as is often the case, there was uncertainty about the very need for adaptations (How badly ailing was the aircraft, really?) as well as uncertainty about the effect and safety of adapting: How much time would the crew have to change their plans? Could they skip fuel dumping and still attempt a landing? Potential adaptations, and the ability to project their potential for success, were not necessarily supported by specific training or overall professional indoctrination. Civil aviation, after all, tends to emphasize model the first model: Stick with procedures and you will most likely be safe (e.g., Lautman & Gallimore, 1987).

Tightening procedural adherence, through threats of punishment or other supervisory interventions, does not remove the double bind. In fact, it may tighten the double bind—making it more difficult for people to develop judgment of how and when to adapt. Increasing the pressure to comply increases the probability of failures to adapt—compelling people to adopt a more conservative response criterion. People will require more evidence for the need to adapt, which takes time, and time may be scarce in cases that call for adaptation (as in the aforementioned case). Merely stressing the importance of following procedures can increase the number of cases in which people fail to adapt in the face of surprise.

Letting people adapt without adequate skill or preparation, on the other hand, can increase the number of failed adaptations. One way out of the double bind is to develop people's skill at adapting. This means giving them the ability to balance the risks between the two possible types of failure: failing to adapt or attempting adaptations that may fail. It requires the development of judgment about local conditions and the opportunities and risks they present, as well as an awareness of larger goals and constraints that operate on the situation. Development of this skill could be construed, to paraphrase Rochlin (1999), as planning for surprise. Indeed, as Rochlin (p. 1549) observed, the culture of safety in high-reliability organizations anticipates and plans for possible failures in "the continuing expectation of future surprise."

Progress on safety also hinges on how an organization responds in the wake of failure (or even the threat of failure). Post-mortems can quickly reveal a gap between procedures and local practice, and hindsight inflates the causal role played by unofficial action (McDonald et al., 2002). The response, then, is often to try to forcibly close the gap between procedures and practice, by issuing more procedures or policing practice more closely. The role of informal patterns of behavior, and what they represent (e.g., resource constraints, organizational deficiencies or managerial ignorance, countervailing goals, peer pressure, professionalism and perhaps even better ways of working) all go misunderstood. Real practice, as done in the vast informal work systems, is driven and kept underground. Even though failures offer each sociotechnical system an opportunity for critical self-examination, accident stories are developed in which procedural deviations play a major, evil role, and are branded as deviant and causal. The official reading of how the system works or is supposed to work is once again re-invented: Rules mean safety, and people should follow them. High-reliability organizations, in contrast, distinguish themselves by their constant investment in trying to monitor and understand the gap between procedures and practice. The common reflex is not to try to close the gap, but to understand why it exists. Such understanding provides insight into the grounds for informal patterns of activity and opens ways to improve safety by sensitivity to people's local operational context.

The Regulator: From Police to Partner

That there is always a tension between centralized guidance and local practice creates a clear dilemma for those tasked with regulating safety-critical industries. The dominant regulatory instrument consists of rules and checking that those rules are followed. But forcing operational people to stick to rules can lead to ineffective, unproductive or even unsafe local actions. For various jobs, following the rules and getting the task done are mutually exclusive. On the other hand, letting people adapt their local practice in the face of pragmatic demands can make them sacrifice global system goals or miss other constraints or vulnerabilities that operate on the system. Helping people solve this fundamental trade-off is not a matter of pushing the criterion one way or the other. Discouraging people's attempts at adaptation can increase the number of failures to adapt in situations where adaptation was necessary. Allowing procedural leeway without encouraging organizations to invest in people's skills at adapting, on the other hand, can increase the number of failed attempts at adaptation.

This means that the gap between rule and task, between written procedure and actual job, needs to be bridged by the regulator as much as by the operator. Inspectors who work for regulators need to apply rules as well: find out what exactly the rules mean and what their implications are when imposed on a field of practice. The development from universalism to particularism applies to regulators too. This raises questions about the role that inspectors should play. Should they function as police—checking to what extent the market is abiding by the laws they are supposed to uphold? In that case, should they apply a black-and-white judgment (which would ground a number of companies immediately)? Or, if there is a gap between procedure and practice that inspectors and operators share and both need to bridge, can inspectors be partners in joint efforts toward progress on safety? The latter role is one that can only develop in good faith, though such good faith may be the very by-product of the development of a new kind of relationship, or partnership, towards progress on safety. Mismatches between rules and practice are no longer seen as the logical conclusion of an inspection, but rather as the starting point, the beginning of joint discoveries about real practice and the context in which it occurs. What are the systemic reasons (organizational, regulatory, resource related) that help create and sustain the mismatch?

The basic criticism of an inspector's role as partner is easy to anticipate: Regulators should not come too close to the ones they regulate, lest their relationship become too cozy and objective judgment of performance against safety criteria become impossible. But regulators need to come close to those they regulate in any case. Regulators (or their inspectors) need to be insiders in the sense of speaking the language of the organization they inspect, understanding the kind of business they are in, in order to gain the respect and credibility of the informants they need most. At the same time, regulators need to be outsiders—resisting getting integrated into the worldview of the one they regulate. Once on the inside of that system and its worldview, it may be increasingly difficult to discover the potential drift into failure. What is normal to the operator is normal to the inspector.

The tension between having to be an insider and an outsider at the same time is difficult to resolve. The conflictual, adversarial model of safety regulation has in many cases not proven productive. It leads to window dressing and posturing on the part of the operator during inspections, and secrecy and obfuscation of safety- and work-related information at all other times. As airline maintenance testifies, real practice is easily driven underground. Even for regulators who apply their power as police rather than as partner, the struggle of having to be insider and outsider at the same time is not automatically resolved. Issues of access to information (the relevant information about how people really do their work, even when the inspector is not there) and inspector credibility, demand that there be a relationship between regulator and operator that allows such access and credibility to develop. Organizations (including regulators) who wish to make progress on safety with procedures need to:

- Monitor the gap between procedure and practice and try to understand why it exists (and resist trying to close it simply telling people to comply).
- Help people develop skills to judge when and how to adapt (and resist only telling people they should follow procedures).

But many organizations or industries do neither. They may not even know, or want to know (or be able to afford to know) about the gap. Take aircraft maintenance again. A variety of workplace factors (communication problems, physical or hierarchical distance, industrial relations) obscure the gap. For example, continued safe outcomes of existing practice give supervisors no reason to question their assumptions about how work is done (if they are safe they must be following procedures down there). There is wider industry ignorance, however (McDonald et al., 2002). In the wake of failure, informal work systems typically retreat from view, gliding out of investigators' reach. What goes misunderstood, or unnoticed, is that informal work systems compensate for the organization's inability to provide the basic resources (e.g., time, tools, documentation with a close relationship to action) needed for task performance. Satisfied that violators got caught and that formal prescriptions of work were once again amplified, the organizational system changes little or nothing. It completes another *cycle of stability*, typified by a stagnation of organizational learning and no progress on safety (McDonald et al.).

GOAL CONFLICTS AND PROCEDURAL DEVIANCE

As discussed in chapter 2, a major engine behind routine divergence from written guidance is the need to pursue multiple goals simultaneously. Multiple goals mean goal conflicts. As Dörner (1989) remarked, "Contradictory goals are the rule, not the exception, in complex situations" (p. 65). In a study of flight dispatchers, for example, Smith (2001) illustrated the basic dilemma. Would bad weather hit a major hub airport or not? What should the dispatchers do with all the airplanes en route? Safety (by making aircraft divert widely around the weather) would be a pursuit that "tolerates a false alarm but deplores a miss" (p. 361). In other words, if safety is the major goal, then making all the airplanes divert even if the weather would not end up at the hub (a false alarm) is much better than not making them divert and sending them headlong into bad weather (a miss). Efficiency, on the other hand, severely discourages the false alarm, whereas it can actually deal with a miss.

As discussed in chapter 2, this is the essence of most operational systems. Though safety is a (stated) priority, these systems do not exist to be safe. They exist to provide a service or product, to achieve economic gain, to maximize capacity utilization. But still they have to be safe. One starting point, then, for understanding a driver behind routine deviations, is to look deeper into these goal interactions, these basic incompatibilities in what people need to strive for in their work. Of particular interest is how people themselves view these conflicts from inside their operational reality, and how this contrasts with management (and regulator) views of the same activities.

NASA's "Faster, Better, Cheaper" organizational philosophy in the late 1990s epitomized how multiple, contradictory goals are simultaneously present and active in complex systems. The loss of the Mars Climate Orbiter and the Mars Polar Lander in 1999 were ascribed in large part to the irreconcilability of the three goals (faster and better and cheaper), which drove down the cost of launches, made for shorter, aggressive mission schedules, eroded personnel skills and peer interaction, limited time, reduced the workforce, and lowered the level of checks and balances normally found (National Aeronautics and Space Administration, 2000). People argued that NASA should pick any two from the three goals. Faster and cheaper would not mean better. Better and cheaper would mean slower. Faster and better would be more expensive. Such reduction, however, obscures the actual reality facing operational personnel in safety-critical settings. These people are there to pursue all three goals simultaneously—fine-tuning their operation, as Starbuck and Milliken (1988) said, to "render it less redundant, more efficient, more profitable, cheaper, or more versatile" (p. 323), fine-tuning, in other words, to make it faster, better, cheaper.

The 2003 Space Shuttle Columbia accident focused attention on the maintenance work that was done on the Shuttle's external fuel tank, once again revealing the differential pressures of having to be safe and getting the job done (better, but also faster and cheaper). A mechanic working for the contractor, whose task it was to apply the insulating foam to the external fuel tank, testified that it took just a couple of weeks to learn how to get the job done, thereby pleasing upper management and meeting production schedules. An older worker soon showed him how he could mix the base chemicals of the foam in a cup and brush it over scratches and gouges in the insulation, without reporting the repair. The mechanic soon found himself doing this hundreds of times, each time without filling out the required paperwork. Scratches and gouges that were brushed over with the mixture from the cup basically did not exist as far as the organization was concerned. And those that did not exist could not hold up the production schedule for the external fuel tanks. Inspectors often did not check. A company program that once had paid workers hundreds of dollars for finding defects had been watered down, virtually inverted by incentives for getting the job done now.

Goal interactions are critical in such experiences, which contain all the ingredients of procedural fluidity, maintenance pressure, the meaning of incidents worth reporting, and their connections to drift into failure. As in most operational work, the distance between formal, externally dictated logics of action and actual work is bridged with the help of those who have been there before, who have learned how to get the job done (without apparent safety consequences), and who are proud to share their professional experience with younger, newer workers. Actual practice by newcomers settles at a distance from the formal description of the job. Deviance becomes routinized. This is part of the vast informal networks characterizing much maintenance work, including informal hierarchies of teachers and apprentices, informal documentation of how to actually get work done, informal procedures and tasks, and informal teaching practices. Inspectors did not check, did not know, or did not report. Managers were happy that production schedules were met and happy that fewer defects were being discovered—normal people doing normal work in a normal organization. Or that is what it seemed to everybody at the time. Once again, the notion of an incident, of something that was worthy of reporting (a defect) got blurred against a background of routine nonconformity. What was normal versus what was deviant was no longer so clear. Goal conflicts between safer, better, and cheaper were reconciled by doing the work more cheaply, superficially better (brushing over gouges), and apparently without cost to safety. As long as orbiters kept coming back safely, the contractor must have been doing something right. Understanding the potential side effects was very difficult given the historical mission success rate. Lack of failures were seen as a validation that current strategies to prevent hazards were sufficient. Could anyone foresee, in a vastly complex system, how local actions as trivial as brushing chemicals from a cup could one day align with other factors to push the system over the edge? Recall from chapter 2: What cannot be believed cannot be seen. Past success was taken as guarantee of continued safety.

The Internalization of External Pressure

Some organizations pass on their goal conflicts to individual practitioners quite openly. Some airlines, for example, pay their crews a bonus for on-time performance. An aviation publication commented on one of those operators (a new airline called Excel, flying from England to holiday destinations): "As part of its punctuality drive, Excel has introduced a bonus scheme to give employees a bonus should they reach the agreed target for the year. The aim of this is to focus everyone's attention on keeping the aircraft on schedule" (Airliner World, 2001, p. 79). Such plain acknowledgment of goal priorities, however, is not common. Most important goal conflicts are never made so explicit, arising rather from multiple irreconcilable directives from different levels and sources, from subtle and tacit pressures, from management or customer reactions to particular trade-offs. Organizations often resort to "conceptual integration, or plainly put, doublespeak" (Dörner, 1989, p. 68). For example, the operating manual of another airline opens by stating that "(1) our flights shall be safe; (2) our flights shall be punctual; (3) our customers will find value for money." Conceptually, this is Dörner's (1989) doublespeak, documentary integration of incompatibles. It is impossible, in principle, to do all three simultaneously, as with NASA's faster, better, cheaper. Whereas incompatible goals arise at the level of an organization and its interaction with its environment, the actual managing of goal conflicts under uncertainty gets pushed down into local operating units—control rooms, cockpits, and the like. There the conflicts are to be negotiated and resolved in the form of thousands of little and larger daily decisions and trade-offs. These are no longer decisions and trade-offs made by the organization, but by individual operators or crews. It is this insidious delegation, this hand-over, where the internalization of external pressure takes place. Crews of one airline describe their ability to negotiate these multiple goals while under the pressure of limited resources as "the blue feeling" (referring to the dominant color of their fleet). This feeling represents the willingness and ability to put in the work to actually deliver on all three goals simultaneously (safety, punctuality, and value for money). This would confirm that practitioners do pursue incompatible goals of faster, better, and cheaper all at the same time and are aware of it too. In fact, practitioners take their ability to reconcile the irreconcilable as a source of considerable professional pride. It is seen as a strong sign of their expertise and competence.

The internalization of external pressure, this usurpation of organizational goal conflicts by individual crews or operators, is not well described or modeled yet. This, again, is a question about the dynamics of the macro–micro connection that we saw in chapter 2. How is it that a global tension between efficiency and safety seeps into local decisions and trade-offs by individual people or groups? These macrostructural forces, which operate on an entire company, find their most prominent expression in how local work groups make assessments about opportunities and risks (see also Vaughan, 1996). Institutional pressures are reproduced, or perhaps really manifested, in what individual people do, not by the organization as a whole. But how does this connection work? Where do external pressures become internal? When do the problems and interests of an organization under pressure of resource scarcity and competition become the problems and interests of individual actors at several levels within that organization?

The connection between external pressure and its internalization is relatively easy to demonstrate when an organization explicitly advertises how operators' pursuit of one goal will lead to individual rewards (a bonus scheme to keep everybody focused on the priority of schedule). But such cases are probably rare, and it is doubtful whether they represent actual internalization of a goal conflict. It becomes more difficult when the connection and the conflicts are more deeply buried in how operators transpose global organizational aims onto individual decisions. For example, the blue feeling signals aircrews' strong identification with their organization (which flies blue aircraft) and what it and its brand stand for (safety, reliability, value for money). Yet it is a feeling that only individuals or crews can have, a feeling because it is internalized. Insiders point out how some crews or commanders have the blue feeling whereas others do not. It is a personal attribute, not an organizational property. Those who do not have the blue feeling are marked by their peers—seldom supervisors—for their insensitivity to, or disinterest in, the multiplicity of goals, for their unwillingness to do substantive cognitive work necessary to reconcile the irreconcilable. These practitioners do not reflect the corps' professional pride because they will always make the easiest goal win over the others (e.g., "Don't worry about customer service or capacity utilization, it's not my job"), choosing the path of least resistance and least work in the eyes of their peers. In the same airline, those who try to adhere to minute rules and regulations are called "Operating Manual worshippers"—a clear signal that their way of dealing with goal contradictions is not only perceived as cognitively cheap (just go back to the book, it will tell you what to do), but as hampering the collective ability to actually get the job done, diluting the blue feeling. The blue feeling, then, is also not just a personal attribute, but an interpeer commodity that affords comparisons, categorizations, and competition among members of the peer group, independent of other layers or levels in the organization. Similar interpeer pride and perception operate as subtle engine behind the negotiation among different goals in other professions too, for example flight dispatchers, air-traffic controllers, or aircraft maintenance workers (McDonald et al., 2002).

The latter group (aircraft maintenance) has incorporated even more internal mechanisms to deal with goal interactions. The demand to meet technical requirements clashes routinely with time or other resource constraints such as inadequate time, personnel, tools, parts, or functional work environment (McDonald et al., 2002). The vast internal, sub-surface networks of routines, illegal documentation, and shortcuts, which from the outside would be seen as massive infringement of existing procedures, are a result of the pressure to reconcile and compromise. Actual work practices constitute the basis for technicians' strong professional pride and sense of responsibility for delivering safe work that exceeds even technical requirements. Seen from the inside, it is the role of the technician to apply judgment founded on his or her knowledge, experience, and skill—not on formal procedure. Those most adept at this are highly valued for their productive capacity even by higher organizational levels. Yet upon formal scrutiny (e.g., an accident inquiry), informal networks and practices often retreat from view, yielding only a bare-bones version of work in which the nature of goal compromises and informal activities is never explicit, acknowledged, understood, or valued. Similar to the British Army on the Somme, management in some maintenance organizations occasionally decides (or pretends) that there is no local confusion, that there are no contradictions or surprises. In their official understanding, there are rules and people who follow the rules, and safe outcomes as a result. People who do not follow the rules are more prone to causing accidents, as the hindsight bias inevitably points out. To people on the work floor, in contrast, management does not even understand the fluctuating pressures on their work, let alone the strategies necessary to accommodate those (McDonald et al.).

Both cases (the blue feeling and maintenance work) challenge human factors' traditional reading of violations as deviant behavior. Human factors wants work to mirror prescriptive task analyses or rules, and violations breach vertical control implemented through such managerial or design directives. Seen from the inside of people's own work, however, violations become compliant behavior. Cultural understandings (e.g., expressed in notions of a blue feeling) affect interpretative work, so that even if people's behavior is objectively deviant, they will see their own conduct as conforming (Vaughan, 1999). Their behavior is compliant with the emerging, local, internalized ways to accommodate multiple goals important to the organization (maximizing capacity utilization but doing so safely, meeting technical requirements, but also deadlines). It is compliant, also, with a complex of peer pressures and professional expectations in which unofficial action yields better, quicker ways to do the job; in which unofficial action is a sign of competence and expertise; where unofficial action can override or outsmart hierarchical control and compensate for higher level organizational deficiencies or ignorance.

ROUTINE NONCONFORMITY

The gap between procedures and practice is not constant. After the creation of new work (e.g., through the introduction of new technology), time can go by before applied practice stabilizes, likely at a distance from the rules as written for the system on the shelf. Social science has characterized this migration from tightly coupled rules to more loosely coupled practice variously as "fine-tuning" (Starbuck & Milliken, 1988) or "practical drift" (Snook, 2000). Through this shift, applied practice becomes the pragmatic imperative; it settles into a system as normative. Deviance (from the original rules) becomes normalized; nonconformity becomes routine (Vaughan, 1996). The literature has identified important ingredients in the normalization of deviance, which can help organizations understand the nature of the gap between procedures and practice:

• Rules that are overdesigned (written for tightly coupled situations, for the worst case) do not match actual work most of the time. In real work, there is slack: time to recover, opportunity to reschedule and get the job done better or more smartly (Starbuck & Milliken). This mismatch creates an inherently unstable situation that generates pressure for change (Snook).

• Emphasis on local efficiency or cost effectiveness pushes operational people to achieve or prioritize one goal or a limited set of goals (e.g., customer service, punctuality, capacity utilization). Such goals are typically easily measurable (e.g., customer satisfaction, on-time performance), whereas it is much more difficult to measure how much is borrowed from safety.

• Past success is taken as guarantee of future safety. Each operational success achieved at incremental distances from the formal, original rules can establish a new norm. From here a subsequent departure is once again only a small incremental step (Vaughan). From the outside, such fine-tuning constitutes incremental experimentation in uncontrolled settings (Starbuck & Milliken)—on the inside, incremental nonconformity is an adaptive response to scarce resources, multiple goals, and often competition.

• Departures from the routine become routine. Seen from the inside of people's own work, violations become compliant behavior. They are compliant with the emerging, local ways to accommodate multiple goals important to the organization (maximizing capacity utilization but doing so safely; meeting technical requirements, but also deadlines). They are compliant, also, with a complex of peer pressures and professional expectations in which unofficial action yields better, quicker ways to do the job; in which unofficial action is a sign of competence and expertise; where unofficial action can override or outsmart hierarchical control and compensate for higher level organizational deficiencies or ignorance.

Although a gap between procedures and practice always exists, there are different interpretations of what this gap means and what to do about it. As pointed out in chapter 6, human factors may see the gap between procedures and practice as a sign of complacency—operators' self-satisfaction with how safe their practice or their system is or a lack of discipline. Psychologists may see routine nonconformity as expressing a fundamental tension between multiple goals (production and safety) that pull workers in opposite directions: getting the job done but also staying safe. Others highlight the disconnect that exists between distant supervision or preparation of the work (as laid down in formal rules) on the one hand, and local, situated action on the other. Sociologists may see in the gap a political lever applied on management by the work floor, overriding or outsmarting hierarchical control and compensating for higher level organizational deficiencies or ignorance. To the ethnographer, routine nonconformity would be interesting not just because of what it says about the work or the work context, but because of what it says about what the work means to the operator.

The distance between procedures and practice can create widely divergent images of work. Is routine nonconformity an expression of elitist operators who consider themselves to be above the law, of people who demonstrate a willingness to ignore the rules? Work in that case is about individual choices, supposedly informed choices between doing that work well or badly, between following the rules or not. Or is routine nonconformity a systematic by-product of the social organization of work, where it emerges from the interactions between organizational environment (scarcity and competition), internalized pressures, and the underspecified nature of written guidance? In that case, work is seen as fundamentally contextualized, constrained by environmental uncertainty and organizational characteristics, and influenced only to a small extent by individual choice. People's ability to balance these various pressures and influences on procedure following depends in large part on their history and experience. And, as Wright and McCarthy (2003) pointed out, there are currently very few ways in which this experience can be given a legitimate voice in the design of procedures.

As chapter 8 shows, a more common way of responding to what is seen as human unreliability is to introduce more automation. Automation has no trouble following algorithms. In fact, it could not run without any. Yet such literalism can be a mixed blessing.

Can We Automate Human Error Out of the System?

If people cannot be counted on to follow procedures, should we not simply marginalize human work? Can automation get rid of human unreliability and error? Automation extends our capabilities in many, if not all, transportation modes. In fact, automation is often presented and implemented precisely because it helps systems and people perform better. It may even make operational lives easier: reducing task load, increasing access to information, helping the prioritization of attention, providing reminders, doing work for us where we cannot. What about reducing human error? Many indeed have the expectation that automation will help reduce human error. Just look at some of the evidence: All kinds of transport achieve higher navigation accuracy with satellite guidance; pilots are now able to circumvent pitfalls such as thunderstorms, windshear, mountains, and collisions with other aircraft; and situation awareness improves dramatically with the introduction of moving map displays.

So with these benefits, can we automate human error out of the system? The thought behind the question is simple. If we automate part of a task, then the human does not carry out that part. And if the human does not carry out that part, there is no possibility of human error. As a result of this logic, there was a time (and in some quarters there perhaps still is) that automating everything we technically could was considered the best idea. The Air Transport Association of America (ATA) observed, for example, that "during the 1970's and early 1980's . . . the concept of automating as much as possible was considered appropriate" (ATA, 1989, p. 4). It would lead to greater safety, greater capabilities, and other benefits.

NEW CAPABILITIES, NEW COMPLEXITIES

But, really, can we automate human error out of the system? There are problems. With new capabilities come new complexities. We cannot just automate part of a task and assume that the human–machine relationship remains unchanged. Though it may have shifted (with the human doing less and the machine doing more), there is still an interface between humans and technology. And the work that goes on at that interface has likely changed drastically. Increasing automation transforms hands-on operators into supervisory controllers, into managers of a suite of automated and other human resources. With their new work come new vulnerabilities, new error opportunities. With of new interfaces (from pointers to pictures, from single parameter gauges to computer displays) come new pathways to human–machine coordination breakdown. Transportation has witnessed the transformation of work by automation first-hand, and documented its consequences widely. Automation does not do away with what we typically call human error, just as (or precisely because) it does not do away with human work. There is still work to do for people. It is not that the same kinds of errors occur in automated systems as in manual systems. Automation changes the expression of expertise and error; it changes how people can perform well and changes how their performance breaks down, if and when it does. Automation also changes opportunities for error recovery (often not for the better) and in many cases delays the visible consequences of errors. New forms of coordination breakdowns and accidents have emerged as a result.

Data Overload

Automation does not replace human work. Instead, it changes the work it is designed to support. And with these changes come new burdens. Take system monitoring, for example. There are concerns that automation can create data overload. Rather than taking away cognitive burdens from people, automation introduces new ones, creating new types of monitoring and memory tasks. Because automation does so much, it also can show much (and indeed, there is much to show). If there is much to show, data overload can occur, especially in pressurized, high-workload, or unusual situations. Our ability to make sense of all the data generated by automation has not kept pace with systems' ability to collect, transmit, transform, and present data.

But data overload is a pretty complex phenomenon, and there are different ways of looking at it (see Woods, Patterson, & Roth, 2002). For example, we can see it as a workload bottleneck problem. When people experience data overload, it is because of fundamental limits in their internal information-processing capabilities. If this is the characterization, then the solution lies in even more automation. More automation, after all, will take work away from people. And taking work away will reduce workload.

One area where the workload-reduction solution to the data-overload problem has been applied is in the design of warning systems. It is there that fears of data overload are often most prominent. Incidents in aviation and other transportation modes keep stressing the need for better support of human problem solving during dynamic fault scenarios. People complain of too much data, of illogical presentations, of warnings that interfere with other work, of a lack of order, and of no rhyme or reason to the way in which warnings are presented. Workload reduction during dynamic fault management is so important because problem solvers in dynamic domains need to diagnose malfunctions while maintaining process integrity. Not only must failures be managed while keeping the process running (e.g., keeping the aircraft flying); their implications for the ability to keep the process running in the first place need to be understood and acted on. Keeping the process intact and diagnosing failures are interwoven cognitive demands in which timely understanding and intervention are often crucial.

A fault in a dynamic processes typically produces a cascade of disturbances or failures. Modern airliners and high-speed vessels have their systems tightly packed together because there is not much room onboard. Systems are also cross-linked in many intricate ways, with electronic interconnections increasingly common as a result of automation and computerization. This means that failures in one system quickly affect other systems, perhaps even along nonfunctional propagation paths. Failure crossover can occur simply because systems are located next to one another, not because they have anything functional in common. This may defy operator logic or knowledge. The status of single components or systems, then, may not be that interesting for a operator. In fact, it may be highly confusing. Rather, the operator must see, through a forest of seemingly disconnected failures, the structure of the problem so that a solution or countermeasure becomes evident. Also, given the dynamic process managed, which issue should be addressed first? What are the postconditions of these failures for the remainder of operations (i.e., what is still operational, how far can I go, what do I need to reconfigure)? Is there any trend? Are there noteworthy events and changes in the monitored process right now? Will any of this get worse? These are the types of questions that are critical to answer in successful dynamic fault management.

Current warning systems in commercial aircraft do not go far in answering these questions, something that is confirmed by pilots' assessments of these systems. For example, pilots comment on too much data, particularly all kinds of secondary and tertiary failures, with no logical order, and primary faults (root causes) that are rarely, if ever, highlighted. The representation is limited to message lists, something that we know hampers operators' visualization of the state of their system during dynamic failure scenarios. Yet not all warning systems are the same. Current warning systems show a range of automated support, from not doing much at all, through prioritizing and sorting warnings, to doing something about the failures, to doing most of the fault management and not showing much at all anymore. Which works best? Is there any merit to seeing data overload as a workload bottleneck problem, and do automated solutions help?

An example of a warning system that basically shows everything that goes wrong inside an aircraft's systems, much in order of appearance, is that of the Boeing 767. Messages are presented chronologically (which may mean the primary fault appears somewhere in the middle or even at the bottom of the list) and failure severity is coded through color. A warning system that departs slightly from this baseline is for example the Saab 2000, which sorts the warnings by inhibiting messages that do not require pilot actions. It displays the remaining warnings chronologically. The primary fault (if known) is placed at the top, however, and if a failure results in an automatic system reconfiguration, then this is shown too. The result is a shorter list than the Boeing's, with a primary fault at the top. Next as an example comes the Airbus A320, which has a fully defined logic for warning-message prioritization. Only one failure is shown at the time, together with immediate action items required of the pilot. Subsystem information can be displayed on demand. Primary faults are thus highlighted, together with guidance on how to deal with them. Finally, there is the MD-11, which has the highest degree of autonomy and can respond to failures without asking the pilot to do so. The only exceptions are nonreversible actions (e.g., an engine shutdown). For most failures, the system informs the pilot of system reconfiguration and presents system status. In addition, the system recognizes combinations of failures and gives a common name to these higher order failures (e.g., Dual Generator).

As could be expected, response latency on the Boeing 767-type warning system is longest (Singer & Dekker, 2000). It takes a while for pilots to sort through the messages and figure out what to do. Interestingly, they also get it wrong more often on this type of system. That is, they misdiagnose the primary failure more often than on any of the other systems. A nonprioritized list of chronological messages about failures seems to defeat even the speed–accuracy trade-off: Longer dwell times on the display do not help people get it right. This is because the production of speed and accuracy are cognitive: Making sense of what is going wrong inside an aircraft's systems is a demanding cognitive task, where problem representation has a profound influence on people's ability to do it successfully (meaning fast and correct). Modest performance gains (faster responses and fewer misdiagnoses) can be seen on a system like that of the Saab 2000, but the

Airbus A320 and MD-11 solutions to the workload bottleneck problem really seem to pay off. Performance benefits really accrue with a system that sorts through the failures, shows them selectively, and guides the pilot in what to do next. In our study, pilots were quickest to identify the primary fault in the failure scenario with such a system, and made no misdiagnoses in assessing what it was (Singer & Dekker). Similarly, a warning system that itself contains or counteracts many of the failures and shows mainly what is left to the pilot seems to help people in quickly identifying the primary fault.

These results, however, should not be seen as justification for simply automating more of the failure-management task. Human performance difficulties associated with high-automation participation in difficult or novel circumstances are well known, such as brittle procedure following where operators follow heuristic cues from the automation rather than actively seeking and dealing with information related to the disturbance chain. Instead, these results indicate how progress can be made by changing the representational quality of warning systems altogether, not just by automating more of the human task portion. If guidance is beneficial, and if knowing what is left is useful, then the results of this study tell designers of warning systems to shift to another view of referents (the thing in the process that the symbol on the display refers to). Warning-system designers would have to get away from relying on single systems and their status as referents to show on the display, and move toward referents that fix on higher order variables that carry more meaning relative to the dynamic fault-management task. Referents could integrate current status with future predictions, for example, or cut across single parameters and individual systems to reveal the structure behind individual failures and show consequences in terms that are operationally immediately meaningful (e.g., loss of pressure, loss of thrust).

Another way of looking at data overload is as a clutter problem—there is simply too much on the display for people to cope with. The solution to data overload as a clutter problem is to remove stuff from the display. In warning-system design, for example, this may result in guidelines that stress how no more than a certain number of lines must be filled up on a warning screen. Seeing data overload as clutter, however, is completely insensitive of context. What seems clutter in one situation may be highly valuable, or even crucial, in another situation. The crash of an Airbus A330 during a test flight at the factory field in Toulouse, France in 1994 provides a good demonstration of this (see Billings, 1997). The aircraft was on a certification test flight to study various pitch-transition control laws and how they worked during an engine failure at low altitude, in a lightweight aircraft with a rearward center of gravity (CG). The flight crew included a highly experienced test pilot, a copilot, a flight-test engineer, and three passengers. Given the lightweight and rearward CG, the aircraft got off the runway quickly and easily and climbed rapidly, with a pitch angle of almost 25° nose-up. The autopilot was engaged 6 seconds after takeoff. Immediately after a short climb, the left engine was brought to idle power and one hydraulic system was shut down in preparation for the flight test. Now the autopilot had to simultaneously manage a very low speed, an extremely high angle of attack, and asymmetrical engine thrust. After the captain disconnected the autopilot (this was only 19 seconds after takeoff) and reduced power on the right engine to regain control of the aircraft, even more airspeed was lost. The aircraft stalled, lost altitude rapidly, and crashed 36 seconds after takeoff.

When the airplane reached a 25° pitch angle, autopilot and flight-director mode information were automatically removed from the primary flight display in front of the pilots. This is a sort of declutter mode. It was found that, because of the high rate of ascent, the autopilot had gone into altitude-acquisition mode (called ALT* in the Airbus) shortly after takeoff. In this mode there is no maximum pitch protection in the autoflight system software (the nose can go as high as the autopilot commands it to go, until the laws of aerodynamics intervene). In this case, at low speed, the autopilot was still trying to acquire the altitude commanded (2,000 feet), pitching up to it, and sacrificing airspeed in the process. But ALT* was not shown to the pilots because of the declutter function. So the lack of pitch protection was not announced, and may not have been known to them. Declutter has not been a fruitful or successful way of trying to solve data overload (see Woods et al., 2002), precisely because of the context problem. Reducing data elements on one display calls for that knowledge to be represented or retrieved elsewhere (people may need to pull it from memory instead), lest it be altogether unavailable.

Merely seeing data overload as a workload or clutter problem is based on false assumptions about how human perception and cognition work. Questions about maximum human data-processing rates are misguided because this maximum, if there is one at all, is highly dependent on many factors, including people's experience, goals, history, and directed attention. As alluded to earlier in the book, people are not passive recipients of observed data; they are active participants in the intertwined processes of observation, action, and sense making. People employ all kinds of strategies to help manage data, and impose meaning on it. For example, they redistribute cognitive work (to other people, to artifacts in the world), they rerepresent problems themselves so that solutions or countermeasures become more obvious. Clutter and workload characterizations treat data as a unitary input phenomenon, but people are not interested in data, they are interested in meaning. And what is meaningful in one situation may not be meaningful in the next. Declutter functions are context insensitive, as are workload-reduction measures. What is interesting, or meaningful, depends on context. This makes designing a warning or display system highly challenging. How can a designer know what the interesting, meaningful or relevant pieces of data will be in a particular context? This takes a deep understanding of the work as it is done, and especially as it will be done once the new technology has been implemented. Recent advances in cognitive work analysis (Vicente, 1999) and cognitive task design (Hollnagel, 2003) presented ways forward, and more is said about such envisioning of future work toward the end of this chapter.

Adapting to Automation, Adapting the Automation

In addition to knowing what (automated) systems are doing, humans are also required to provide the automation with data about the world. They need to input things. In fact, one role for people in automated systems is to bridge the context gap. Computers are dumb and dutiful: They will do what they are programmed to do, but their access to context, to a wider environment, is limited—limited, in fact, to what has been predesigned or preprogrammed into them. They are literalist in how they work. This means that people have to jump in to fill a gap: They have to bridge the gulf between what the automation knows (or can know) and what really is happening or relevant out there in the world. The automation, for example, will calculate an optimal descent profile in order to save as much fuel as possible. But the resulting descent may be too steep for crew (and passenger) taste, so pilots program in an extra tailwind, tricking the computers into descending earlier and eventually more shallow (because the tailwind is fictitious). The automation does not know about this context (preference for certain descent rates over others), so the human has to bridge the gap. Such tailoring of tools is a very human thing to do: People will shape tools to fit the exact task they must fulfill. But tailoring is not risk- or problem-free. It can create additional memory burdens, impose cognitive load when people cannot afford it, and open up new error opportunities and pathways to coordination breakdowns between human and machine.

Automation changes the task for which it was designed. Automation, though introducing new capabilities, can increase task demands and create new complexities. Many of these effects are in fact unintended by the designers. Also, many of these side effects remain buried in actual practice and are hardly visible to those who only look for the successes of new machinery. Operators who are responsible for (safe) outcomes of their work are known to adapt technology so that it fits their actual task demands. Operators are known to tailor their working strategies so as to insulate themselves from the potential hazards associated with using the technology. This means that the real effects of technology change can remain hidden beneath a smooth layer of adaptive performance. Operational people will make it work, no matter how recalcitrant or ill suited to the domain the automation, and its operating procedures, really may be. Of course, the occasional breakthroughs in the form of surprising accidents provide a window onto the real nature of automation and its operational consequences. But such potential lessons quickly glide out of view under the pressure of the fundamental surprise fallacy.

Apparently successful adaptation by people in automated systems, though adaptation in unanticipated ways, can be seen elsewhere in how pilots deal with automated cockpits. One important issue on high-tech flight decks is knowing what mode the automation is in (this goes for other applications such as ship's bridges too: Recall the Royal Majesty from chap. 5). Mode confusion can lie at the root of automation surprises, with people thinking that they told the automation to do one thing whereas it was actually doing another. How do pilots keep track of modes in an automated cockpit? The formal instrument for tracking and checking mode changes and status is the FMA, or flight-mode annunciator, a small strip that displays contractions or abbreviations of modes (e.g., Heading Select mode is shown as HDG or HDG SEL) in various colors, depending on whether the mode is armed (i.e., about to become engaged) or engaged. Most airline procedures require pilots to call out the mode changes they see on the FMA.

One study monitored flight crews during a dozen return flights between Amsterdam and London on a full flight simulator (Björklund, Alfredsson, & Dekker, 2003). Where both pilots were looking and how long was measured by EPOG (eye-point-of-gaze) equipment, which uses different kinds of techniques ranging from laser beams to measuring and calibrating saccades, or eye jumps that can track the exact focal point of a pilot's eyes in a defined visual field (see Fig. 8.1).

Pilots do not look at the FMA much at all. And they talk even less about it. Very few call-outs are made the way they should be (according to the procedures). Yet this does not seem to have an effect on automation-mode awareness, nor on the airplane's flight path. Without looking or talking, most pilots apparently still know what is going on inside the automation. In this one study, 521 mode changes occurred during the 12 flights. About 60% of these were pilot induced (i.e., because of the pilot changing a setting in the automation), the rest were automation induced. Two out of five mode changes were never visually verified (meaning neither pilot looked at their FMA during 40% of all mode changes). The pilot flying checked a little less than the pilots not flying, which could be a natural reflection of the role division: Pilots who are flying the aircraft have other sources of flight-related data they need to look at, whereas the pilot not flying can oversee the entire process, thereby engaging more often in checks of what the automation modes are. There are also differences between captains and first officers as well (even after you correct for pilot-flying vs. pilot-not-flying

FIG. 8.1.   Example of pilot EPOG (eye point of gaze) fixations on a primary flight display (PFD) and map display in an automated cockpit. The top part of the PFD is the flight-mode annunciator (FMA; Björklund et al., 2003).

roles). Captains visually verified the transitions in 72% of the cases, versus 47% for first officers. This may mirror the ultimate responsibility that captains have for safety of flight, yet there was no expectation that this would translate into such concrete differences in automation monitoring. Amount of experience on automated aircraft types was ruled out as being responsible for the difference.

Of 512 mode changes, 146 were called out. If that does not seem like much, consider this: Only 32 mode changes (that is about 6%) were called out after the pilot looked at the FMA. The remaining call-outs came either before looking at the FMA, or without looking at the FMA at all. Such a disconnect between seeing and saying suggests that there are other cues that pilots use to establish what the automation is doing. The FMA does not serve as a major trigger for getting pilots to call out modes. Two out of five mode transitions on the FMA are never even seen by entire flight crews. In contrast to instrument monitoring in nonglass-cockpit aircraft, monitoring for mode transitions is based more on a pilot's mental model of the automation (which drives expectations of where and when to look) and an understanding of what the current situation calls for. Such models are often incomplete and buggy and it is not surprising that many mode transitions are neither visually nor verbally verified by flight crews.

At the same time, a substantial number of mode transitions are actually anticipated correctly by flight crews. In those cases where pilots do call out a mode change, four out of five visual identifications of those mode changes are accompanied or preceded by a verbalization of their occurrence. This suggests that there are multiple, underinvestigated resources that pilots rely

on for anticipating and tracking automation-mode behavior (including pilot mental models). The FMA, designed as the main source of knowledge
about automation status, actually does not provide a lot of that knowledge.
It triggers a mere one out of five call-outs, and gets ignored altogether by
entire crews for a whole 40% of all mode transitions. Proposals for new regulations are unfortunately taking shape around the same old display concepts. For example, Joint Advisory Circular ACJ 25.1329 (Joint Aviation Authorities, 2003, p. 28) said that: "The transition from an armed mode to an
engaged mode should provide an additional attention-getting feature, such
as boxing and flashing on an electronic display (per AMJ25-11) for a suitable, but brief, period (e.g., ten seconds) to assist in flight crew awareness."
But flight-mode annunciators are not at all attention getting, whether there
is boxing or flashing or not. Indeed, empirical data show (as it has before,
see Mumaw, Sarter, & Wickens, 2001) that the FMA does not "assist in flight
crew awareness" in any dominant or relevant way. If design really is to capture crew's attention about automation status and behavior, it will have to
do radically better than annunciating abstruse codes in various hues and
boxing or flashing times.

The call-out procedure appears to be miscalibrated with respect to real
work in a real cockpit, because pilots basically do not follow formal verification and call-out procedures at all. Forcing pilots to visually verify the FMA
first and then call out what they see bears no similarity to how actual work is
done, nor does it have much sensitivity to the conditions under which such
work occurs. Call-outs may well be the first task to go out the window when
workload goes up, which is also confirmed by this type of research. In addition to the few formal call-outs that do occur, pilots communicate implicitly
and informally about mode changes. Implicit communication surrounding
altitude capture could for example be "Coming up to one-three-zero, (capture)" (referring to flight level 130). There appear to be many different
strategies to support mode awareness, and very few of them actually overlap
with formal procedures for visual verification and call-outs. Even during the
12 flights of the Björklund et al. (2003) study, there were at least 18 different strategies that mixed checks, timing, and participation. These strategies
seem to work as well as, or even better than, the official procedure, as crew
communications on the 12 flights revealed no automation surprises that
could be traced to a lack of mode awareness. Perhaps mode awareness does
not matter that much for safety after all.

There is an interesting experimental side effect here: If mode awareness
is measured mainly by visual verification and verbal call-outs, and crews neither look nor talk, then are they unaware of modes, or are the researchers
unaware of pilots' awareness? This poses a puzzle: Crews who neither talk
nor look can still be aware of the mode their automation is in, and this, indeed seems to be the case. But how, in that case, is the researcher (or your

company, or line-check pilot) to know? The situation is one answer. By simply looking at where the aircraft is going, and whether this overlaps with the pilots' intentions, an observer can get to know something about apparent pilot awareness. It will show whether pilots missed something or not. In the research reported here however, pilots missed nothing: There were no unexpected aircraft behaviors from their perspective (Björklund et al., 2003). This can still mean that the crews were either not aware of the modes and it did not matter, or they were aware but the research did not capture it. Both may be true.

MABA-MABA OR ABRACADABRA

The diversity of experiences and research results from automated cockpits shows that automation creates new capabilities and complexities in ways that may be difficult to anticipate. People adapt to automation in many different ways, many of which have little resemblance to formally established procedures for interacting with the automation. Can automation, in a very Cartesian, dualistic sense, replace human work, thereby reducing human error? Or is there a more complex coevolution of people and technology? Engineers and others involved in automation development are often led to believe that there is a simple answer, and in fact a simple way of getting the answer. MABA-MABA lists, or "Men-Are-Better-At, Machines-Are-Better-At" lists have appeared over the decades in various guises. What these lists basically do is try to enumerate the areas of machine and human strengths and weaknesses, in order to provide engineers with some guidance on which functions to automate and which ones to give to the human. The process of function allocation as guided by such lists sounds straightforward, but is actually fraught with difficulty and often unexamined assumptions.

One problem is that the level of granularity of functions to be considered for function allocation is arbitrary. For example, it depends on the model of information processing on which the MABA-MABA method is based (Hollnagel, 1999). In Parasuraman, Sheridan, and Wickens (2000), four stages of information processing (acquisition, analysis, selection, response) form the guiding principle to which functions should be kept or given away, but this is an essentially arbitrary decomposition based on a notion of a human–machine ensemble that resembles a linear input–output device. In cases where it is not a model of information processing that determines the categories of functions to be swapped between human and machine, the technology itself often determines it (Hollnagel, 1999). MABA-MABA attributes are then cast in mechanistic terms, derived from technological metaphors. For example, Fitts (1951) applied terms such as information capacity and computation in his list of attributes for both the human and the machine. If the technology gets to pick the battlefield (i.e., determine the language of attributes) it will win most of them back for itself. This results in human-uncentered systems where typically heuristic and adaptive human abilities such as not focusing on irrelevant data, scheduling and reallocating activities to meet current constraints, anticipating events, making generalizations and inferences, learning from past experience, and collaborating (Hollnagel) easily fall by the wayside.

Moreover, MABA-MABA lists rely on a presumption of fixed human and machine strengths and weaknesses. The idea is that, if you get rid of the (human) weaknesses and capitalize on the (machine) strengths, you will end up with a safer system. This is what Hollnagel (1999) called "function allocation by substitution." The idea is that automation can be introduced as a straightforward substitution of machines for people—preserving the basic system while improving some of its output measures (lower workload, better economy, fewer errors, higher accuracy, etc.). Indeed, Parasuraman et al. (2000) recently defined automation in this sense: "Automation refers to the full or partial replacement of a function previously carried out by the human operator" (p. 287). But automation is more than replacement (although perhaps automation is about replacement from the perspective of the engineer). The really interesting issues from a human performance standpoint emerge after such replacement has taken place.

Behind the idea of substitution lies the idea that people and computers (or any other machines) have fixed strengths and weaknesses and that the point of automation is to capitalize on the strengths while eliminating or compensating for the weaknesses. The problem is that capitalizing on some strength of computers does not replace a human weakness. It creates new human strengths and weaknesses—often in unanticipated ways (Bainbridge, 1987). For instance, the automation strength to carry out long sequences of action in predetermined ways without performance degradation amplifies classic human vigilance problems. It also exacerbates the system's reliance on the human strength to deal with the parametrization problem, or literalism (automation does not have access to all relevant world parameters for accurate problem solving in all possible contexts). As we have seen, however, human efforts to deal with automation literalism, by bridging the context gap, may be difficult because computer systems can be hard to direct (How do I get it to understand? How do I get it to do what I want?). In addition, allocating a particular function does not absorb this function into the system without further consequences. It creates new functions for the other partner in the human–machine equation—functions that did not exist before, for example, typing, or searching for the right display page, or remembering entry codes. The quest for a priori function allocation, in other words, is intractable (Hollnagel & Woods, 1983), and not only this: Such new kinds of work create new error opportunities (What was that code again? Why can't I find the right page?).

TRANSFORMATION AND ADAPTATION

Automation produces qualitative shifts. Automating something is not just a matter of changing a single variable in an otherwise stable system (Woods & Dekker, 2001). Automation transforms people's practice and forces them to adapt in novel ways: "It alters what is already going on—the everyday practices and concerns of a community of people—and leads to a resettling into new practices" (Flores, Graves, Hartfield, & Winograd, 1988, p. 154). Unanticipated consequences are the result of these much more profound, qualitative shifts. For example, during the Gulf War in the early 1990s, "almost without exception, technology did not meet the goal of unencumbering the personnel operating the equipment. Systems often required exceptional human expertise, commitment, and endurance" (Cordesman & Wagner, 1996, p. 25).

Where automation is introduced, new human roles emerge. Engineers, given their professional focus, may believe that automation transforms the tools available to people, who will then have to adapt to these new tools. In chapter 9 we see how, according to some researchers, the removal of paper flight-progress strips in air-traffic control represents a transformation of the workplace, to which controllers only need to adapt (they will compensate for the lack of flight progress strips). In reality, however, people's practice gets transformed by the introduction of new tools. New technology, in turn, gets adapted by people in locally pragmatic ways so that it will fit the constraints and demands of actual practice. For example, controlling without flight-progress strips (relying more on the indications presented on the radar screen) asks controllers to develop and refine new ways of managing airspace complexity and dynamics. In other words, it is not the technology that gets transformed and the people who adapt. Rather, people's practice gets transformed and they in turn adapt the technology to fit their local demands and constraints.

The key is to accept that automation will transform people's practice and to be prepared to learn from these transformations as they happen. This is by now a common (but not often successful) starting point in contextual design. Here the main focus of system design is not the creation of artifacts per se, but getting to understand the nature of human practice in a particular domain, and changing those work practices rather than just adding new technology or replacing human work with machine work. This recognizes that:

- Design concepts represent hypotheses or beliefs about the relationship between technology and human cognition and collaboration.
- They need to subject these beliefs to empirical jeopardy by a search for disconfirming and confirming evidence.
- These beliefs about what would be useful have to be tentative and open to revision as they learn more about the mutual shaping that goes on between artifacts and actors in a field of practice.

Subjecting design concepts to such scrutiny can be difficult. Traditional validation and verification techniques applied to design prototypes may turn up nothing, but not necessarily because there is nothing that could turn up. Validation and verification studies typically try to capture small, narrow outcomes by subjecting a limited version of a system to a limited test. The results can be informative, but hardly about the processes of transformation (different work, new cognitive and coordination demands) and adaptation (novel work strategies, tailoring of the technology) that will determine the sources of a system's success and potential for failure once it has been fielded. Another problem is that validation and verification studies need a reasonably ready design in order to carry any meaning. This presents a dilemma: By the time results are available, so much commitment (financial, psychological, organizational, political) has been sunk into the particular design that any changes quickly become unfeasible.

Such constraints through commitment can be avoided if human factors can say meaningful things early on in a design process. What if the system of interest has not been designed or fielded yet? Are there ways in which we can anticipate whether automation, and the human role changes it implies, will create new error problems rather than simply solving old ones? This has been described as Newell's catch: In order for human factors to say meaningful things about a new design, the design needs to be all but finished. Although data can then be generated, they are no longer of use, because the design is basically locked. No changes as a result of the insight created by human factors data are possible anymore. Are there ways around this catch? Can human factors say meaningful things about a design that is nowhere near finished? One way that has been developed is future incident studies, and the concept they have been tested on is exception management.

AUTOMATION AND EXCEPTION MANAGEMENT

One role that may fit the human well is that of exception manager. Introducing automation to turn people into exception managers can sound like a good idea. In ever busier systems, where operators are vulnerable to problems of data overload, turning humans into exception managers is a power-fully attractive concept. It has, for example, been practiced in the dark cockpit design that essentially keeps the human operator out of the loop (all the annunciator lights are out in normal operating conditions) until something interesting happens, which may then be the time for the human to intervene. This same envisioned role, of exception manager, dominates recent ideas about how to effectively let humans control ever increasing air-traffic loads. Perhaps, the thought goes, controllers should no longer be in charge of all the parameters of every flight in their sector. A core argument is that the human controller is a limiting factor in traffic growth. Too many aircraft under one single controller leads to memory overload and the risk of human error. Decoupling controllers from all individual flights in their sectors, through greater computerization and automation on the ground and greater autonomy in the air, is assumed to be the way around this limit.

The reason we may think that human controllers will make good exception managers is that humans can handle the unpredictable situations that machines cannot. In fact, this is often a reason why humans are still to be found in automated systems in the first place (see Bainbridge, 1987). Following this logic, controllers would be very useful in the role of traffic manager, waiting for problems to occur in a kind of standby mode. The view of controller practice is one of passive observer, ready to act when necessary.

But intervening effectively from a position of disinvolvement has proven to be difficult—particularly in air-traffic control. For example, Endsley, Mogford, Allendoerfer, and Stein (1997) pointed out, in a study of direct routings that allowed aircraft deviations without negotiations, that with more freedom of action being granted to individual aircraft, it became more difficult for controllers to keep up with traffic. Controllers were less able to predict how traffic patterns would evolve over a foreseeable timeframe. In other studies too, passive monitors of traffic seemed to have trouble maintaining a sufficient understanding of the traffic under their control (Galster, Duley, Masolanis, & Parasuraman, 1999), and were more likely to overlook separation infringements (Metzger & Parasuraman, 1999). In one study, controllers effectively gave up control over an aircraft with communication problems, leaving it to other aircraft and their collision-avoidance systems to sort it out among themselves (Dekker & Woods, 1999). This turned out to be the controllers' only route out of a fundamental double bind: If they intervened early they would create a lot of workload problems for themselves (suddenly a large number of previously autonomous aircraft would be under their control). Yet if they waited on intervention (in order to gather more evidence on the aircraft's intentions), they would also end up with an unmanageable workload and very little time to solve anything in. Controller disinvolvement can create more work rather than less, and produce a greater error potential.

This brings out one problem of envisioning practice, of anticipating how automation will create new human roles and what the performance consequences of those roles will be. Just saying "manager of exceptions" is insufficient: It does not make explicit what it means to practice. What work does an exception manager do? What cues does he or she base decisions on? The downside of underspecification is the risk of remaining trapped in a disconnected, shallow, unrealistic view of work. And when our view of (future) practice is disconnected from many of the pressures, challenges, and constraints operating in that world, our view of practice is distorted from the beginning. It misses how operational people's strategies are often intricately adapted to deal effectively with these constraints and pressures.

There is an upside to underspecification, however, and that is the freedom to explore new possibilities and new ways to relax and recombine the multiple constraints, all in order to innovate and improve. Will automation help you get rid of human error? With air-traffic controllers as exception managers, it is interesting to think about how the various designable objects would be able to support them in exception management. For example, visions of future air-traffic control systems typically include data linking as an advance that avoids the narrow bandwidth problem of voice communications—thus enhancing system capacity. In one study (Dekker & Woods, 1999), a communications failure affected an aircraft that had also suffered problems with its altitude reporting (equipment that tells controllers how high it is and whether it is climbing or descending). At the same time, this aircraft was headed for streams of crossing air traffic. Nobody knew exactly how datalink, another piece of technology not connected to altitude-encoding equipment, would be implemented (its envisioned use was, and is, to an extent underspecified). One controller, involved in the study, had the freedom to suggest that air-traffic control should contact the airline's dispatch or maintenance office to see whether the aircraft was climbing or descending or level. After all, data link could be used by maintenance and dispatch personnel to monitor the operational and mechanical status of an aircraft, so "if dispatch monitors power settings, they could tell us," the controller suggested. Others objected because of the coordination overheads this would create. The ensuing discussion showed that, in thinking about future systems and their consequences for human error, we can capitalize on underspecification if we look for the so-called leverage points (in the example: data link and other resources in the system) and a sensitivity to the fact that envisioned objects only become tools through use—imagined or real (data links to dispatch become a backup air-traffic control tool).

Anticipating the consequences of automation on human roles is also difficult because—without a concrete system to test—there are always multiple versions of how the proposed changes will affect the field of practice in the future. Different stakeholders (in air-traffic control this would be air carriers, pilots, dispatchers, air-traffic controllers, supervisors, flow controllers) have different perspectives on the impact of new technology on the nature of practice. The downside of this plurality is a kind of parochialism where people mistake their partial, narrow view for the dominant view of the future of practice, and are unaware of the plurality of views across stakeholders. For example, one pilot claimed that greater autonomy for airspace users is "safe, period" (Baiada, 1995). The upside of plurality is the triangulation that is possible when the multiple views are brought together. In examining the relationships, overlaps, and gaps across multiple perspectives, we are better able to cope with the inherent uncertainty built into looking into the future.

A number of future incident studies (see Dekker & Woods, 1999) examined controllers' anomaly response in future air-traffic control worlds precisely by capitalizing on this plurality. To study anomaly response under envisioned conditions, groups of practitioners (controllers, pilots, and dispatchers) were trained on proposed future rules. They were brought together to try to apply these rules in solving difficult future airspace problems that were presented to them in several scenarios. These included aircraft decompression and emergency descents, clear air turbulence, frontal thunderstorms, corner-post overloading (too many aircraft going to one entry point for airport area), and priority air-to-air refueling and consequent airspace restrictions and communication failures. These challenges, interestingly, were largely rule or technology independent: They can happen in airspace systems of any generation. The point was not to test the anomaly response performance of one group against that of another, but to use triangulation of multiple stakeholder viewpoints—anchored in the task details of a concrete problem—to discover where the envisioned system would crack, where it would break down. Validity in such studies derives from: (a) the extent to which problems to be solved in the test situation represent the vulnerabilities and challenges that exist in the target world, and (b) the way in which real problem-solving expertise is brought to bear by the study participants.

Developers of future air-traffic control architectures have been envisioning a number of predefined situations that call for controller intervention, a kind of reasoning that is typical for engineering-driven decisions about automated systems. In air-traffic management, for example, potentially dangerous aircraft maneuvers, local traffic density (which would require some density index), or other conditions that compromise safety would make it necessary for a controller to intervene. Such rules, however, do not reduce uncertainty about whether to intervene. They are all a form of threshold crossing—intervention is called for when a certain dynamic density has been reached or a number of separation miles has been transgressed. But threshold-crossing alarms are very hard to get right—they come either too early or too late. If too early, a controller will lose interest in them: The alarm will be deemed alarmist. If the alarm comes too late, its contribution to flagging or solving the problem will be useless and it will be deemed incompetent. The way in which problems in complex, dynamic worlds grow and escalate, and the nature of collaborative interactions, indicate that recognizing exceptions in how others (either machines or people) are handling anomalies is complex. The disappointing history of automating problem diagnosis inspires little further hope. Threshold-crossing alarms cannot make up for a disinvolvement—they can only make a controller acutely aware of those situations in which it would have been nice to have been involved from the start.

Future incident studies allow us to extend the empirical and theoretical base on automation and human performance. For example, supervisory-control literature makes no distinction between anomalies and exceptions. This indistinction results from the source of supervisory-control work: How do people control processes over physical distances (time lag, lack of access, etc.). However, air-traffic control augments the issue of supervisory control with a cognitive distance: Airspace participants have some system knowledge and operational perspective, as do controllers, but there are only partial overlaps and many gaps. Studies on exception management in future air-traffic control force us to make a distinction between anomalies in the process, and exceptions from the point of view of the supervisor (controller). Exceptions can arise in cases where airspace participants are dealing with anomalies (e.g., an aircraft with pressurization or communications problems) in a way that forces the controller to intervene. An exception is a judgement about how well others are handling or going to handle disturbances in the process. Are airspace participants handling things well? Are they going to get themselves in trouble in the future? Judging whether airspace users are going to get in trouble in their dealings with a process disturbance would require a controller to recognize and trace a situation over time—contradicting arguments that human controllers make good stand-by interveners.

Will Only the Predicted Consequences Occur?

In developing new systems, it is easy for us to become miscalibrated. It is easy for us to become overconfident that if our envisioned system can be realized, the predicted consequences and only the predicted consequence will occur. We lose sight of the fact that our views of the future are tentative hypotheses and that we would actually need to remain open to revision, that we need to continually subject these hypotheses to empirical jeopardy.

One way to fool ourselves into thinking that only the predicted consequences will occur when we introduce automation is to stick with substi-
tutional practice of function allocation. Substitution assumes a fundamentally uncooperative system architecture in which the interface between
human and machine has been reduced to a straightforward "you do this, I
do that" trade. If that is what it is, of course we should be able to predict the
consequences. But it is not that simple. The question for successful automation is not who has control over what or how much. That only looks at the
first parts, the engineering parts. We need to look beyond this and start asking humans and automation the question: "How do we get along together?"
Indeed, where we really need guidance today is in how to support the coordination between people and automation. In complex, dynamic, nonde-
terministic worlds, people will continue to be involved in the operation of
highly automated systems. The key to a successful future of these systems
lies in how they support cooperation with their human operators—not only
in foreseeable standard situations, but also under novel, unexpected circumstances.

One way to frame the question is how to turn automated systems into effective team players (Sarter & Woods, 1997). Good team players make their
activities observable to fellow team players, and are easy to direct. To be observable, automation activities should be presented in ways that capitalize
on well-documented human strengths (our perceptual system's acuity to
contrast, change and events, our ability to recognize patterns and know
how to act on the basis of this recognition, e.g., Klein). For example:

- Event based: Representations need to highlight changes and events in
  ways that the current generation of state-oriented displays do not.
- Future oriented: In addition to historical information, human operators in dynamic systems need support for anticipating changes and
  knowing what to expect and where to look next.
- Pattern based: Operators must be able to quickly scan displays and pick
  up possible abnormalities without having to engage in difficult cognitive work (calculations, integrations, extrapolations of disparate pieces
  of data). By relying on pattern- or form-based representations, automation has an enormous potential to convert arduous mental tasks into
  straightforward perceptual ones.

Team players are directable when the human operator can easily and efficiently tell them what to do. Designers could borrow inspiration from how
practitioners successfully direct other practitioners to take over work.
These are intermediate, cooperative modes of system operation that allow

human supervisors to delegate suitable subproblems to the automation, just as they would be delegated to human crew members. The point is not to make automation into a passive adjunct to the human operator who then needs to micromanage the system each step of the way. This would be a waste of resources, both human and machine. Human operators must be allowed to preserve their strategic role in managing system resources as they see fit, given the circumstances.

Chapter 9

Will the System Be Safe?

How do you know whether a new system will be safe? As chapter 8 showed, automating parts of human work may make a system safer, but they may not. The Alaska Airlines 261 accident discussed in chapter 2 illustrates how difficult it is to know whether a system is going to be safe during its operational lifetime. In the case of the DC-9 trim system, bridging the gap between producing a system and running it proved quite difficult. Certifying that the system was safe, or airworthy, when it rolled out of the factory with zero flying hours was one thing. Certifying that it would stay safe during a projected lifetime proved to be quite another. Alaska 261 shows how large the gulf between making a system and maintaining it can be.

The same is true for sociotechnical systems. Take the issue of flight-progress strips in air-traffic control. The flight strip is a small paper slip with flight-plan data about each controlled aircraft's route, speed, altitude, times over waypoints, and other characteristics (see Fig. 9.1). It is used by air-traffic controllers in conjunction with a radar representation of air traffic. A number of control centers around the world are doing away with these paper strips, to replace them with automated flight-tracking systems. Each of these efforts requires, in principle, a rigorous certification process. Different teams of people look at color coding, letter size and legibility, issues of human–computer interaction, software reliability and stability, seating arrangements, button sensitivities, and so forth, and can spend a decade following the footsteps of a design process to probe and poke it with methods and forms and questionnaires and tests and checklists and tools and guidelines—all in an effort to ensure that local human factors or ergonomics standards have been met. But such static

S KA 6337 7351	320	310	OSY 1936	FUR 1945	TEN 1952

FIG. 9.1.   Example of a flight strip. From left to right it shows the airplane's flight number and transponder code, the entry altitude of the aircraft into the controller's sector (FL320), the exit level (FL310), and what times it is expected to fly across particular waypoints along its route in the controller's sector.

snapshots may mean little. A lineup of microcertificates of usability does not guarantee safety. As soon as they hit the field of practice, systems start to drift. A year (or a month) after its inception, no sociotechnical system is the same as it was in the beginning. As soon as a new technology is introduced, the human, operational, organizational system that is supposed to make the technology work forces it into locally practical adaptations. Practices (procedures, rules) adapt around the new technology, and the technology in turn is reworked, revised, and amended in response to the emergence of practical experience.

THE LIMITS OF SAFETY CERTIFICATION

System safety is more than the sum of the certified parts. A redundant torque tube inside of a jackscrew, for example, does nothing to maintain the integrity of a DC-9 trim system without a maintenance program that guarantees continued operability. But ensuring the existence of such a maintenance system is nothing like understanding how the local rationality of such a system can be sustained (we're doing the right thing, the safe thing) while safety standards are in fact continually being eroded (e.g., from 350- to 2,550-hour lubrication interval). The redundant components may have been built and certified. The maintenance program (with 2,550-hour lubrication intervals—certified) may be in place. But safe parts do not guarantee system safety.

Certification processes do not typically take lifetime wear of parts into account when judging an aircraft airworthy, even if such wear will render an aircraft, like Alaska 261, quite unworthy of flying. Certification processes certainly do not know how to take sociotechnical adaptation of new equipment, and the consequent potential for drift into failure, into account when looking at nascent technologies. Systemic adaptation or wear is not a criterion in certification decisions, nor is there a requirement to put in place an organization to prevent or cover for anticipated wear rates or pragmatic adaptation, or fine-tuning. As a certification engineer from the regulator testified, "Wear is not considered as a mode of failure for either a system safety analysis or for structural considerations" (NTSB, 2002, p. 24). Because how do you take wear into account? How can you even predict with any accuracy how much wear will occur? McDonnell-Douglas surely had it wrong when it anticipated wear rates on the trim jackscrew assembly of its DC-9. Originally, the assembly was designed for a service life of 30,000 flight hours without any periodic inspections for wear. But within a year, excessive wear had been discovered nonetheless, prompting a reconsideration.

The problem of certifying a system as safe to use can become even more complicated if the system to be certified is sociotechnical and thereby even less calculable. What does wear mean when the system is sociotechnical rather than consisting of pieces of hardware? In both cases, safety certification should be a lifetime effort, not a still assessment of decomposed system status at the dawn of a nascent technology. Safety certification should be sensitive to the coevolution of technology and its use, its adaptation. Using the growing knowledge base on technology and organizational failure, safety certification could aim for a better understanding of the ecology in which technology is released—the pressures, resource constraints, uncertainties, emerging uses, fine-tuning, and indeed lifetime wear.

Safety certification is not just about seeing whether components meet criteria, even if that is what it often practically boils down to. Safety certification is about anticipating the future. Safety certification is about bridging the gap between a piece of gleaming new technology in the hand now, and its adapted, coevolved, grimy, greased-down wear and use further down the line. But we are not very good at anticipating the future. Certification practices and techniques oriented toward assessing the standard of current components do not translate well into understanding total system behavior in the future. Making claims about the future, then, often hangs on things other than proving the worthiness of individual parts. Take the trim system of the DC-9 again.

The jackscrew in the trim assembly had been classified as a "structure" in the 1960s, leading to different certification requirements from when it would have been seen as a system. The same piece of hardware, in other words, could be looked at as two entirely different things: a system, or a structure. In being judged a structure, it did not have to undergo the required system safety analysis (which may, in the end, still not have picked up on the problem of wear and the risks it implied). The distinction, this partition of a single piece of hardware into different lexical labels, however, shows that airworthiness is not a rational product of engineering calculation. Certification can have much more to do with localized engineering judgments, with argument and persuasion, with discourse and renaming, with the translation of numbers into opinion, and opinion into numbers—all of it based on uncertain knowledge.

As a result, airworthiness is an artificially binary black-or-white verdict (a jet is either airworthy or it is not) that gets imposed on a very grey, vague, uncertain world—a world where the effects of releasing a new technology into actual operational life are surprisingly unpredictable and incalculable. Dichotomous, hard yes or no meets squishy reality and never quite gets a genuine grip. A jet that was judged airworthy, or certified as safe, may or may not be in actual fact. It may be a little bit unairworthy. Is it still airworthy with an end-play check of .0042 inches, the set limit? But "set" on the basis of what? Engineering judgment? Argument? Best guess? Calculations? What if a following end-play check is more favorable? The end-play check itself is not very reliable. The jet may be airworthy today, but no longer tomorrow (when the jackscrew snaps). But who would know?

The pursuit of answers to such questions can precede or accompany certification efforts. Research, that putatively objective scientific encounter with empirical reality, can assist in the creation of knowledge about the future, as shown in chapter 8. So what about working without paper flight strips? The research community has come to no consensus on whether air traffic control can actually do without them, and if it does, how it succeeds in keeping air-traffic under control. Research results are inconclusive. Some literature suggested that flight strips are expendable without consequences for safety (e.g., Albright, Truitt, Barile, Vortac, & Manning, 1996), whereas others argued that air-traffic control is basically impossible without them (e.g., Hughes, Randall, & Shapiro, 1993). Certification guidance that could be extracted from the research base can go either way: It is either safe or unsafe to do away with the flight strips, depending on whom you listen to. What matters most for credibility is whether the researcher can make statements about human work that a certifier can apply to the coming, future use of a system. In this, researchers appeared to rely on argument and rhetoric, as much as on method, to justify that the results they found are applicable to the future.

LEIPZIG AS LEGITIMATE

For human factors, the traditionally legitimate way of verifying the safety of new technology is to conduct experiments in the laboratory. Say that researchers want to test whether operators can safely use voice-input systems, or whether their interpretation of some target is better on three-dimensional displays. The typical strategy is to build microversions of the future system and expose a limited number of participants to various conditions, some or all of which may contain partial representations of a target system. Through its controlled settings, laboratory research already makes some sort of verifiable step into the future. Empirical contact with a world to be designed is ensured because some version of that future world has been prefabricated in the lab. This also leads to problems. Experimental steps into the future are necessarily narrow, which affects the generalizability of research findings. The mapping between test and target situations may miss several important factors.

In part as a result of a restricted integration of context, laboratory studies can yield divergent and eventually inconclusive results. Laboratory research on decision making (Sanders & McCormick, 1997), for example, has found several biases in how decision makers deal with information presented to them. Can new technology circumvent the detrimental aspects of such biases, which, according to some views, would lead to human error and safety problems? One bias is that humans are generally conservative and do not extract as much information from sources as they optimally should. Another bias, derived from the same experimental research, is that people have a tendency to seek far more information than they can absorb adequately. Such biases would seem to be in direct opposition to each other. It means that reliable predictions of human performance in a future system may be difficult to make on the basis of such research. Indeed, laboratory findings often come with qualifying labels that limit their applicability. Sanders and McCormick (1997), for example, advised: "When interpreting the . . . findings and conclusions, keep in mind that much of the literature is comprised of laboratory studies using young, healthy males doing relatively unmotivating tasks. The extent to which we can generalize to the general working population is open to question" (p. 572).

Whether the question remains open does not seem to matter. Experimental human factors research in the laboratory holds a special appeal because it makes mind measurable, and it even allows mathematics to be applied to the results. Quantitativism is good: It helps equate psychology with natural science, shielding it from the unreliable wanderings through mental life using dubious methods like introspection. The large-scale university laboratories that are now a mainstay of many human factors departments were a 19th-century European invention, pioneered by scientists such as the chemist Justus Liebig. Wundt of course started the trend in psychology with his Leipzig laboratory (see chap. 5). Leipzig did psychology a great service: Psychophysics and its methods of inquiry introduced psychology as a serious science, as something realist, with numbers, calculations, and equations. The systematization, mechanization, and quantification of psychological research in Leipzig, however, must be seen as an antimovement against earlier introspection and rationalism.

Echoes of Leipzig still sound loudly today. A quantitativist preference remains strong in human factors. Empiricist appeals (the pursuit of real measurable facts through experiment) and a strong reliance on Cartesian–Newtonian interpretations of natural science equal to those of, say, physics, may help human factors retain credibility in a world of constructed hardware and engineering science, where it alone dabbles in the fuzziness of psychology. In a way, then, quantitativist human factors or engineering psychology is still largely the sort of antimovement that Wundt formed with his Leipzig laboratory. It finds its expression in a pursuit of numbers and statistics, lest engineering consumers of the research results (and their government or other sponsors) suspect the results to be subjective and untrustworthy.

The quantification and mechanization of mind and method in human factors are good only because they are not something else (i.e., foggy rationalism or unreliable introspection), not because they are inherently good or epistemologically automatically justifiable. The experimental method is good for what it is not, not for what it is. One can see this in the fact that quantitative research in mainstream human factors never has to justify its method (that method is good because at least it is not that other, vague stuff). Qualitative research, on the other hand, is routinely dismissed as insufficiently empirical and will always be required to justify its method. Anything perceived to be sliding toward rationalism, subjectivism, and nonsystematic introspection is highly suspicious, not because it is, but because of what it evokes: a fear that human factors will be branded unscientific. Now these fears are nothing new. They have inspired many a split or departure in the history of psychology. Recall Watson's main concern when launching behaviorism. It was to rescue psychology from vague subjectivist introspection (by which he even meant Wundt's systematic, experimental laboratory research) and plant it firmly within the natural science tradition. Ever since Newton read the riot act on what scientific was to be, psychology and human factors have struggled to find an acceptance and an acceptability within that conceptualization.

Misconceptions About the Qualitative–Quantitative Relationship

Whether quantitative or qualitative research can make more valid claims about the future (thereby helping in the certification of a system as safe to use) is contested. At first sight, qualitative, or field studies, are about the present (otherwise there is no field to study). Quantitative research may test actual future systems, but the setting is typically so contrived and limited that its relationship to a real future is tenuous. As many have pointed out, the difference between quantitative and qualitative research is actually not so great (e.g., Woods, 1993; Xiao & Vicente, 2000). Claims of epistemological privilege by either are counterproductive, and difficult to substantiate. A method becomes superior only if it better helps researchers answer the question they are pursuing, and in this sense, of course, the differences between qualitative and quantitative research can be real. But dismissing qualitative work as subjective misses the point of quantitative work. Squeezing numbers out of an experimental encounter with reality, and then closing the gap to a concept-dependent conclusion on what you just saw, requires generous helpings of interpretation. As we see in the following discussion, there is a great deal of subjectivism in endowing numbers with meaning. Moreover, seeing qualitative inquiry as a mere protoscientific prelude to real quantitative research misconstrues the relationship and overestimates quantitative work. A common notion is that qualitative work should precede quantitative research by generating hypotheses that can then be tested in more restricted settings. This may be one relationship. But often quantitative work only reveals the how or what (or how much) of a particular phenomenon. Numbers in themselves can have a hard time revealing the why of the phenomenon. In this case, quantitative work is the prelude to real qualitative research: It is experimental number crunching that precedes and triggers the study of meaning.

Finally, a common claim is that qualitative work is high in external validity and low in internal validity. Quantitative research, on the other hand, is thought to be low in external validity and high in internal validity. This is often used as justification for either approach and it must rank among the most misconstrued arguments in scientific method. The idea is that internal validity is high because experimental laboratory research allows an investigator almost full control over the conditions in which data are gathered. If the experimenter did not make it happen, either it did not happen, or the experimenter knows about it, so that it can be dealt with as a confound. But the degree of control in research is often overestimated. Laboratory settings are simply another kind of contextualized setting, in which all kinds of subtle influences (social expectations, people's life histories) enter and influence performance just like they would in any other contextualized setting. The degree of control in qualitative research, on the other hand, is often simply assumed to be low. And much qualitative work indeed adds to that image. But rigor and control is definitely possible in qualitative work: There are many ways in which a researcher can become confident about systematic relationships between different factors. Subjectivism in interpretation is not more necessary in qualitative than in quantitative research. Qualitative work, on the other hand, is not automatically externally valid simply because it takes place in a field (applied) setting. Each encounter with empirical reality, whether qualitative or quantitative, generates context-specific data—data from that time and place, from those people, in that language—that are by definition nonexportable to other settings. The researcher has to engage in analysis of those data in order to bring them up to a concept-dependent level, from which terms and conclusions can be taken to other settings.

The examples that follow play out these issues. But the account is about more than the real or imagined opposition between qualitative and quantitative work. The question is how human factors research, quantitative or qualitative, can contribute to knowing whether a system will be safe to use.

EXPERIMENTAL HUMAN FACTORS RESEARCH
ON FLIGHT STRIPS: AN EXAMPLE

One way to find out if controllers can control air-traffic without the aid of flight strips is to test it in an experimental setting. You take a limited number of controllers, and put them through a short range of tasks to see how they do. In their experiments, Albright et al. (1996) deployed a wide array of measurements to find out if controllers perform just as well in a condition with no strips as in a condition with strips. The work they performed was part of an effort by the U.S. Federal Aviation Administration, a regulator (and ultimately the certifier of any future air-traffic control system in the U.S.). In their study, the existing air-traffic control system was retained, but to compare stripped versus stripless control, the researchers removed the flight strips in one condition:

> The first set of measurements consisted of the following: total time watching the PVD [plan view display, or radar screen], number of FPR [flight plan requests], number of route displays, number of J-rings used, number of conflict alerts activated, mean time to grant pilot requests, number of unable requests, number of requests ignored, number of controller-to-pilot requests, number of controller-to-center requests, and total actions remaining to complete at the end of the scenario. (Albright et al., p. 6)

The assumption that drives most experimental research is that reality (in this case about the use and usefulness of flight strips) is objective and that it can be discovered by the researcher wielding the right measuring instruments. This is consistent with the structuralism and realism of human factors. The more measurements, the better, the more numbers, the more you know. This is assumed to be valid even when an underlying model that would couple the various measurements together into a coherent account of expert performance is often lacking (as it is in Albright et al., 1996, but also in many folk models in human factors). In experimental work, the number and diversity of measurements can become the proxy indicator of the accuracy of the findings, and of the strength of the epistemological claim (Q: So how do you know what you know? A: Well, we measured this, and this, and this, and that, and . . .). The assumption is that, with enough quantifiable data, knowledge can eventually be offered that produces an accurate and definitive account of a particular system. More of the same will eventually lead to something different. The strong influence that engineering has had on human factors (Batteau, 2001) makes this appear as just common sense. In engineering, technical debates are closed by amassing results from tests and experience; the essence of the craft is to convert uncertainty into certainty. Degrees of freedom are closed through numbers; ambiguity is worked out through numbers; uncertainty is reduced through numbers (Vaughan, 1996).

Independent of the number of measurements, each empirical encounter is of necessity limited, in both place and time. In the case of Albright et al. (1996), 20 air-traffic controllers participated in two simulated airspace conditions (one with strips and one without strips) for 25 minutes each. One of the results was that controllers took longer to grant pilot requests when they did not have access to flight strips, presumably because they had to assemble the basis for a decision on the request from other information sources. The finding is anomalous compared to other results, which showed no significant difference between workload and ability to keep control over the traffic situation across the strip–no strip conditions, leading to the conclusion that "the presence or absence of strips had no effect on either performance or perceived workload. Apparently, the compensatory behaviors were sufficient to maintain effective control at what controllers perceived to be a comparable workload" (Albright et al., p. 11). Albright et al. explained the anomaly as follows: "Since the scenarios were only 25 minutes in length, controllers may not have had the opportunity to formulate strategies about how to work without flight strips, possibly contributing to the delay" (p. 11).

At a different level, this explanation of an anomalous datum implies that the correspondence between the experimental setting and a future system and setting may be weak. Lacking a real chance to learn how to formulate strategies for controlling traffic without flight strips, it would be interesting to pursue the question of how controllers in fact remained in control over the traffic situation and kept their workload down. It is not clear how this lack of a developed strategy can affect the number of requests granted but not the perceived workload or control performance. Certifiers may, or perhaps should, wonder what 25 minutes of undocumented struggle tells them about a future system that will replace decades of accumulated practice. The emergence of new work and establishment of new strategies is a fundamental accompaniment to the introduction of new technology, representing a transformation of tasks, roles, and responsibilities. These shifts are not something that could easily be noticed within the confines of an experimental study, even if controllers were studied for much longer than 25 minutes. Albright et al. (1996), resolved this by placing the findings of control performance and workload earlier in their text: "Neither performance nor perceived workload (as we measured them in this study) was affected when the strips were removed" (p. 8). The qualification that pulled the authority of the results back into the limited time and place of the experimental encounter (how we measured them in this study), were presented parenthetically and thus accorded less central importance (Golden-Biddle & Locke, 1993). The resulting qualification suggests that comparable performance and workload may be mere artifacts of the way the study was conducted, of how these things were measured at that time and place, with those tools, by those researchers. The qualification, however, was in the middle of the paper, in the middle of a paragraph, and surrounded by other paragraphs adorned with statistical allusions. Nothing of the qualification remained at the end of the paper, where the conclusions presented these localized findings as universally applicable truths.

Rhetoric, in other words, is enlisted to deal with problematic areas of epistemological substance. The transition from localized findings (in this study the researchers found no difference in workload or performance the way they measured them with these 20 controllers) to generalizable principles (we can do away with flight strips) essentially represents a leap of faith. As such, central points of the argument were left unsaid or were difficult for the reader to track, follow, or verify. By bracketing doubt this way, Albright et al. (1996) communicated that there was nothing, really, to doubt. Authority (i.e., true or accurate knowledge) derives from the replicable, quantifiable experimental approach. As Xiao and Vicente (2000) argued, it is very common for quantitative human factors research not to spend much time on the epistemological foundation of its work. Most often it moves unreflectively from a particular context (e.g., an experiment) to concepts (not having strips is safe), from data to conclusions, or from the modeled to the model. The ultimate resolution of the fundamental constraint on empirical work (i.e., each empirical encounter is limited to a time and place) is that more research is always necessary. This is regarded as a highly reasonable conclusion of most quantitative human factors, or indeed any, experimental work. For example, in the Albright et al. study, one constraint was the 25-minute time limit on the scenarios played. Does flight-strip removal actually change controller strategies in ways that were not captured by the present study? This would seem to be a key question. But again, the reservation was bracketed. Whether or not the study answered this question does not in the end weaken the study's main conclusion: "(Additional research is necessary to determine if there are more substantial long term effects to strip removal)" (p. 12).

In addition, the empirical encounter of the Albright et al. (1996) study was limited because it only explored one group of controllers (upper airspace). The argument for more research was drafted into service for legitimizing (not calling into question) results of the study: "Additional studies should be conducted with field controllers responsible for other types of sectors (e.g., low altitude arrival, or non-radar) to determine when, or if, controllers can compensate as successfully as they were able to in the current investigation" (p. 12). The idea is that more of the same, eventually, will lead to something different, that a series of similar studies over time will produce a knowledge increment useful to the literature and useful to the consumers of the research (certifiers in this case). This, once again, is largely taken for granted in the human factors community. Findings will invariably get better next time, and such successive, incremental enhancement is a legitimate route to the logical human factors end point: the discovery of an objective truth about a particular human–machine system and, through this, the revelation of whether it will be safe to use or not.

Experimental work relies on the production of quantifiable data. Some of this quantification (with statistical ornaments such as F-values and standard deviations) was achieved in Albright et al. (1996) by converting tickmarks on lines of a questionnaire (called the "PEQ," or post-experimental questionnaire) into an ordinal series of digits:

> The form listed all factors with a 9.6 centimeter horizontal line next to each. The line was marked low on the left end and high on the right end. In addition, a vertical mark in the center of the line signified the halfway mark. The controllers were instructed to place an X on the line adjacent to the factor to indicate a response. . . . The PEQ scales were scored by measuring distance from the right anchor to the mark placed by the controller on a horizontal line (in centimeters). . . . Individual repeated measures ANOVAs [were then conducted]. (pp. 5–8)

The veneration of numbers in this case, however, went a step too far. ANOVAs cannot be used for the kind of data gathered through PEQ scales. The PEQ is made up of so-called ordinal scales. In ordinal scales, data categories are mutually exclusive (a tickmark cannot be at two distances at the same time), they have some logical order, and they are scored according to the amount of a particular characteristic they possess (in this case, distance in centimeters from the left anchor). Ordinal scales, however, do not represent equal differences (a distance of 2 cm does not represent twice as much of the category measured as a distance of 1 cm), as interval and ratio scales do.

Besides, reducing complex categories such as "usefulness" or "likeability" to distances along a few lines probably misses out on an interesting ideographic reality beneath all of the tickmarks. Put in experimental terms, the operationalization of usefulness as the distance from a tickmark along a line is not particularly high on internal validity. How can the researcher be sure that usefulness means the same thing to all responding controllers? If different respondents have different ideas of what usefulness meant during their particular experimental scenario, and if different respondents have different ideas of how much usefulness a tickmark, say, in the middle of the line represents, then the whole affair is deeply confounded. Researchers do not know what they are asking and do not know what they are getting in reply. Further numeric analysis is dealing with apples and oranges. This is one of the greater risks of folk modeling in human factors. It assumes that everybody understands what usefulness means, and that everybody has the same definition. But these are generous and untested assumptions. It was only with qualitative inquiry that researchers could ensure that there was some consensus on understandings of usefulness with respect to the controlling task with or without strips. Or they could discover that there was no consensus and then control for it. This would be one way to deal with the confound.

It may not matter, and it may not have been noticed. Numbers are good. Also, the linear, predictable format of research writing, as well as the use of abbreviated statistical curios throughout the results section, represent a rhetoric that endows the experimental approach with its authority—authority in the sense of privileged access to a particular layer or slice of empirical reality that others outside the laboratory setting do or do not have admittance to. Other rhetoric invented particularly for the study (e.g., PEQ scales for questions presented to participants after their trials in Albright et al., 1996) certifies the researchers' unique knowledge of this slice of reality. It validates the researcher's competence to tell readers what is really going on there. It may dissuade second-guessing. Empirical results are deemed accurate by virtue of a controlled encounter, a standard reporting format that shows logical progress to objective truths and statements (introduction, method, results, discussion, and summary), and an authoritative dialect intelligible only to certified insiders.

Closing the Gap to the Future

Because of some limited correspondence between the experiment and the system to be designed, quantitative research seemingly automatically closes the gap to the future. The stripless condition in the research (even if contrived by simply leaving out one artifact [the flight strip] from the present) is a model of the future. It is an impoverished model to be sure, and one that offers only a partial window onto what future practice and performance may be like (despite the epistemological reservations about the authenticity of that future discussed earlier). The message from Albright et al.'s (1996) encounter with the future is that controllers can compensate for the lack of flight strips. Take flight strips away, and controllers compensate for the lack of information by seeking information elsewhere (the radar screen, flight-plan readouts, controller-to-pilot requests). Someone might point out that Albright et al. prejudged the use and usefulness of flight strips in the first few sentences of their introduction, that they did not see their data as an opportunity to seek alternative interpretations: "Currently, en route control of high altitude flights between airports depends on two primary tools: the computer-augmented radar information available on the Plan View Display (PVD) and the flight information available on the Flight Progress Strip" (p. 1). This is not really an enabling of knowledge, it is the imposition of it. Here, flight strips are not seen as a problematic core category of controller work, whose use and usefulness would be open to negotiation, disagreement, or multiple interpretations. Instead, flight strips function as information-retrieval devices. Framed as such, the data and the argument can really only go one way: By removing one source of information, controllers will redirect their information-retrieving strategies onto other devices and sources. This displacement is possible, it may even be desirable, and it is probably safe: "Complete removal of the strip information and its accompanying strip marking responsibilities resulted in controllers compensating by retrieving information from the computer" (Albright et al., p. 11). For a certifier, this closes a gap to the future: Removing one source of information will result in people finding the information elsewhere (while showing no decrement in performance or increment in workload). The road to automation is open and people will adapt successfully, for that has been scientifically proven. Therefore, doing away with the flight strips is (probably) safe, and certifiable as such.

If flight strips are removed, then what other sources of information should remain available? Albright et al. (1996) inquired about what kind of information controllers would minimally like to preserve: Route of flight scored high, as did altitude information and aircraft call sign. Naming these categories gives developers the opportunity to envision an automated version of the flight strip that presents the same data in digital format, one that substitutes a computer-based format for the paper-based one, without any consequences for controller performance. Such a substitution, however, may overlook critical factors associated with flight strips that contribute to safe practice, and that would not be incorporated or possible in a computerized version (Mackay, 2000).

Any signs of potential ambiguity or ambivalence about what else flight strips may mean to those working with them were not given further consideration beyond a brief mention in the experimental research write-up—not because these signs were actively, consciously stifled, but because they were inevitably deleted as Albright et al. (1996) carried out and wrote up their encounter with empirical reality. Albright et al. explicitly solicited qualitative, richer data from their participants by asking if controllers themselves felt that the lack of strips impaired their performance. Various controllers indicated how strips help them preplan and that, without strips, they cannot preplan. The researchers, however, never unpacked the notion of preplanning or investigated the role of flight strips in it. Again, such notions (e.g., preplanning) are assumed to speak for themselves, taken to be self-evident. They require no deconstruction, no further interpretive work. Paying more attention to these qualitative responses could create noise that confounds experimental accuracy. Comments that preplanning without strips was impossible hinted at flight strips as a deeper, problematic category of controller work. But if strips mean different things to different controllers, or worse, if preplanning with strips means different things to different controllers, then the experimental bedrock of comparing comparable people across comparable conditions would disappear. This challenges in a profound way the nomothetic averaging out of individual differences. Where individual differences are the nemesis of experimental research, interpretive ambiguity can call into question the legitimacy of the objective scientific enterprise.

QUALITATIVE RESEARCH ON FLIGHT STRIPS: AN EXAMPLE

Rather than looking at people's work from the outside in (as do quantitative experiments), qualitative research tries to understand people's work from the inside out. When taking the perspective of the one doing the work, how does the world look through his or her eyes? What role do tools play for people themselves in the accomplishments of their tasks; how do tools affect their expression of expertise? An interpretive perspective is based on the assumption that people give meaning to their work and that they can express those meanings through language and action. Qualitative research interprets the ways in which people make sense of their work experiences by examining the meanings that people use and construct in light of their situation (Golden-Biddle & Locke, 1993).

The criteria and end points for good qualitative research are different than from those in quantitative research. As a research goal, accuracy is practically and theoretically unobtainable. Qualitative research is relentlessly empirical, but it rarely achieves finality in its findings. Not that quantitative research ever achieves finality (remember that virtually every experimental report finishes with the exhortation that more research is necessary). But qualitative researchers admit that there is never one accurate description or analysis of a system in question, no definitive account—only versions. What flight strips exactly do for controllers is forever subject to interpretation; it will never be answered objectively or finitely, never be closed to further inquiry. What makes a version good, though, or credible, or worth paying attention to by a certifier, is its authenticity. The researcher has to not only convince the certifier of a genuine field experience in writing up the research account, but also make intelligible what went on there. Validation from outside the field emerges from an engagement with the literature (What have others said about similar contexts?) and from interpretation (How well are theory and evidence used to make sense of this particular context?). Field research, though critical to the ethnographic community as a stamp of authenticity, is not necessarily the only legitimate way to generate qualitative data. Surveys of user populations can also be tools that support qualitative inquiry.

Find Out What the Users Think

The reason that qualitative research may appeal to certifiers is that it lets the informants, the users, speak—not through the lens of an experiment, but on the users' terms and initiative. Yet this is also where a central problem lies. Simply letting users speak can be of little use. Qualitative research is not (or should not be) plain conversational mappings—a direct transfer from field setting to research account. If human factors would (or continues to) practice and think about ethnography in these terms, doubts about both the method and the data it yields will continue to surface. What certifiers, as consumers of human factors research, care about is not what users say in raw, unpacked form, but about what their remarks mean for work, and especially for future work. As Hughes et al. (1993) put it: "It is not that users cannot talk about what it is they know, how things are done, but it needs bringing out and directing toward the concerns of the design itself" (p. 138). Within the human factors community, qualitative research seldom takes this extra step. What human factors requires is a strong ethnography, one that actually makes the hard analytical move from user statements to a design language targeted at the future.

A massive qualitative undertaking related to flight strips was the Lancaster University project (Hughes et al., 1993). Many man-months were spent (an index of the authenticity of the research) observing and documenting air-traffic control with flight strips. During this time the researchers developed an understanding of flight strips as an artifact whose functions derive from the controlling work itself. Both information and annotations on the strip and the active organization of strips among and between controllers were essential: "The strip is a public document for the members of the (controlling) team; a working representation of an aircraft's control history and a work site of controlling. Moving the strips is to organize the information in terms of work activities and, through this, accomplishing the work of organizing the traffic" (Hughes et al., pp. 132–133). Terms such as *working representation* and *organizing traffic* are concepts, or categories, that were abstracted well away from the masses of deeply context-specific field notes and observations gathered in the months of research. Few controllers would themselves use the term working representation to explain what flight strips mean to them. This is good. Conceptual abstraction allows a researcher to reach a level of greater generality and increased generalizability (see Woods, 1993; Xiao & Vicente, 2000). Indeed, working representation may be a category that can lead to the future, where a designer would be looking to computerize a working representation of flight information, and a certifier would be evaluating whether such a computerized tool is safe to use. But such higher order interpretive work is seldom found in human factors research. It would separate ethnography and ethnographic argument from research that simply makes claims based on authenticity. Even Hughes et al. (1993) relied on authenticity alone when they told of the various annotations made on flight strips, and did little more than parrot their informants:

> Amendments may be done by the controller, by the chief, or less often, by one of the "wings." "Attention-getting" information may also be written on the strips, such as arrows indicating unusual routes, symbols designating "crossers, joiners and leavers" (that is, aircraft crossing, leaving or joining the major traffic streams), circles around unusual destinations, and so on. (p. 132)

Though serving as evidence of socialization, of familiarity and intimacy, speaking insider language is not enough. By itself it is not helpful to certifiers who may be struggling with evaluating a version of air-traffic control without paper flight strips. Appeals to authenticity ("Look, I was there, and I understand what the users say") and appeals to future relevance ("Look, this is what you should pay attention to in the future system") can thus pull in opposite directions: the former toward more the context specific that is hardly generalizable, the latter toward abstracted categories of work that can be mapped onto yet-to-be-fielded future systems and conceptions of work. The burden to resolve the tension should not be on the certifier or the designer of the system, it should be on the researcher. Hughes et al. (1993) agreed that this bridge-building role should be the researcher's:

> Ethnography can serve as another bridge between the users and the designers. In our case, controllers have advised on the design of the display tool with the ethnographer, as someone knowledgeable about but distanced from the work, and, on the one hand able to appreciate the significance of the controllers' remarks for their design implications and, on the other hand, familiar enough with the design problems to relate them to the controllers' experiences and comments. (p. 138)

Hostage to the Present, Mute About the Future

Hughes et al. (1993) research account actually missed the "significance of controller remarks for their design implications" (p. 138). No safety implications were extracted. Instead the researchers used insider language to forward insider opinions, leaving user statements unpacked and largely underanalyzed. Ethnography essentially gets confused with what informants say and consumers of the research are left to pick and choose among the statements. This is a particularly naive form of ethnography, where what informants can tell researchers is equated or confused with what strong, analytical ethnography (and ethnographic argument) could reveal. Hughes et al. relied on informant statements to the extent they did because of a common belief that the work that their informants did, and the foundational categories that informed it are for the most part self-evident; close to what we would regard as common sense. As such, they require little, if any, analytic effort to discover. It is an ethnography reduced to a kind of mediated user show-and-tell for certifiers—not as thorough analysis of the foundational categories of work. For example, Hughes et al concluded that "(flight strips) are an essential feature of 'getting the picture,' 'organising the traffic,' which is the means of achieving the orderliness of the traffic" (p. 133).

So flight strips help controllers get the picture. This kind of statement is obvious to controllers and merely repeats what everyone already knows. If ethnographic analysis cannot go beyond common sense, it merely privileges the status quo. As such, it offers certifiers no way out: A system without flight strips would not be safe, so forget it. There is no way for a certifier to circumvent the logical conclusion of Hughes et al. (1993): "The importance of the strip to the controlling process is difficult to overestimate" (p. 133). So is it safe? Going back to Hughes et al.: "For us, such questions were not easily answerable by reference to work which is as subtle and complex as our ethnographic analysis had shown controlling to be" (p. 135).

Such surrender to the complexity and intricacy of a particular phenomenon is consistent with what Dawkins (1986, p. 38) called the "argument from personal incredulity." When faced with highly complicated machinery or phenomena, it is easy to take cover behind our own sense of extreme wonder, and resist efforts at explanation. In the case of Hughes et al. (1993), it recalls an earlier reservation: "The rich, highly detailed, highly textured, but nevertheless partial and selective descriptions associated with ethnography would seem to contribute little to resolving the designers problem where the objective is to determine what should be designed and how" (p. 127).

Such justification ("It really is too complex and subtle to communicate to you") maneuvers the entire ethnographic enterprise out of the certifier's view as something not particularly helpful. Synthesizing the complexity and subtlety of a setting should not be the burden of the certifier. Instead, this is the role of the researcher; it is the essence of strong ethnography. That a phenomenon is remarkable does not mean it is inexplicable; so if we are unable to explain it, "we should hesitate to draw any grandiose conclusions from the fact of our own inability" (Dawkins, 1986, p. 39).

Informant remarks such as "Flight strips help me get the mental picture" should serve as a starting point for qualitative research, not as its conclusion. But how can researchers move from native category to analytic sense? Qualitative work should be hermeneutic and circular in nature: not aiming for a definitive description of the target system, but rather a continuous reinterpretation and reproblematization of the successive layers of data mined from the field. Data demand analysis. Analysis in turn guides the search for more data, which in turn demand further analysis: Categories are continually revised to capture the researcher's (and, hand in hand, the practitioner's) evolving understanding of work. There is a constant interplay between data, concepts, and theory.

The analysis and revision of categories is a hallmark of strong ethnography, and Ross's (1995) study of flight-progress strips in Australia serves as an interesting example. Qualitative in nature, Ross's research relied on surveys of controllers using flight strips in their current work. Surveys are often derided by qualitative researchers for imposing the researcher's understanding of the work onto the data, instead of the other way around (Hughes et al., 1993). Demonstrating that it is not just the empirical encounter or rhetorical appeals to authenticity that matter (through large numbers of experimental probes or months of close observation), the survey results Ross gathered were analyzed, coded, categorized, recoded and recategorized until the inchoate masses of context-specific controller remarks began to form sensible, generalizable wholes that could meaningfully speak to certifiers.

Following previous categorizations of flight-strip work (Della Rocco, Manning, & Wing, 1990), Ross (1995) moves down from these conceptual descriptions of controller work and up again from the context-specific details, leaving several layers of intermediate steps. In line with characterizations of epistemological analysis through abstraction hierarchies (see Xiao & Vicente, 2000), each step from the bottom up is more abstract than the previous one; each is cast less in domain-bound terms and more in concept-dependent terms than the one before. Beyer and Holtzblatt (1998) referred to this process as *induction*: reasoning from the particular to the general. One example from Ross (p. 27) concerns domain-specific controller activities such as "entering a pilot report; composing a flight plan amendment." These lower level, context-specific data are of course not without semantic load themselves: it is always possible to ask further questions and descend deeper into the world of meanings that these simple, routine activities have for the people who carry them out. Indeed, we have to ask if we can only go up from the context-specific level—maintained in human factors as the most atomistic, basic, low-level data set (see Woods, 1993). In Ross's data, researchers should still question the common sense behind the otherwise taken-for-granted entering of a pilot report: What does a pilot report mean for the controller in a particular context (e.g., weather related), what does entering this report mean for the controller's ability to manage other traffic issues in the near future (e.g., avoiding sending aircraft into severe turbulence)?

While alluding to even more fine-grained details and questions later, these types of activities also point to an intentional strategy at a higher level of analysis (Della Rocco et al., 1990): that of the "transformation or translation of information for entry into the system," which, at an even higher level of analysis, could be grouped under a label coding, together with other such strategies (Ross, 1995, p. 27). Part of this coding is symbolic, in that it uses highly condensed markings on flight strips (underlining, black circles, strike-throughs) to denote and represent for controllers, what is going on. The highly intricate nature of even one flight (where it crosses vs. where it had planned to cross a sector boundary, what height it will be leaving when, whether it has yet contacted another frequency, etc.) can be collapsed or amortized by simple symbolic notation—one line or circle around a code on the strip that stands for a complex, multidimensional problematic that other controllers can easily recognize. Unable to keep all the details of what a flight would do stable in the head, the controller compresses complexity, or amortizes it, as Hollan, Hutchins, and Kirsh (2000) would say, by letting one symbol stand for complex concepts and interrelationships, some even temporal.

Similarly, "recognizing a symbol for a handoff" (on a flight strip), though allowing further unpacking (e.g., what do you mean "recognize"?), is an instance of a tactic that "transforms or translates information received," which in turn represents a larger controller competency of "decoding," which in its turn is also part of a strategy to use symbolic notation to collapse or amortize complexity (Ross, 1995, p. 27). From recognizing a symbol for a hand-off to the collapsing of complexity, there are four steps, each more abstract and less in domain terms than the one before. Not only do these steps allow others to assess the analytical work for its worth, but the destination of such induction is actually a description of work that can be used for guiding the evaluation of a future system. Inspired by Ross's analysis, we can surmise that controllers rely on flight strips for:

- Amortizing or collapsing complexity (what symbolic notation conveys).

- Supporting coordination (who gets which flight strip next from whom).
- Anticipating dynamics (how much is to come, from where, when, in what order).

These (no longer so large) jumps to the highest level of abstraction can now be made—identifying the role the flight strip has in making sense of workplace and task complexity. Although not so much a leap of faith any longer (because there are various layers of abstraction in between), the final step, up to the highest level conceptual description, still appears to hold a certain amount of creative magic. Ross (1995) revealed little of the mechanisms that actually drive his analysis. There is no extensive record that tracks the transformation of survey data into conceptual understandings of work. Perhaps these transformations are taken for granted too: The mystery is left unpacked because it is assumed to be no mystery. The very process by which the researcher manages to migrate from user-language descriptions of daily activities to conceptual languages less anchored in the present, remains largely hidden from view. No ethnographic literature guides specifically the kinds of inferences that can be drawn up to the highest level of conceptual understanding. At this point, a lot of leeway is given (and reliance placed on) the researcher and his or her (keenly) developed insight into what activities in the field really mean or do for people who carry them out. The problems of this final step are known and acknowledged in the qualitative research community. Vaughan (1996) and other sociologists referred to it as making the macro–micro connection: locating general meaning systems (e.g., symbolic notation, off-loading) in local contexts (placing a circle around a set of digits on the flight strip). Geertz (1973) noted how inferences that try to make the macro–micro connection often resemble "perfected impressionism" in which "much has been touched but little grasped" (p. 312). Such inferences tend to be evocative, resting on suggestion and insinuation more than on analysis (Vaughan, 1996).

In qualitative research, lower levels of analysis or understanding always underconstrain the inferences that can be drawn further on the way to higher levels (see Hollan et al., 2000). At each step, alternative interpretations are possible. Qualitative work does not arrive at a finite description of the system or phenomenon studied (nor does quantitative research, really). But qualitative work does not even aim or pretend to do so (Batteau, 2001). Results are forever open to further interpretation, forever subject to increased problematization. The main criterion, therefore, to which we should hold the inferences drawn is not accuracy (Golden-Biddle & Locke, 1993), but plausibility: Does the conceptual description make sense—especially to the informants, to the people who actually do the work? This also motivates the continuous, circular nature of qualitative analysis: reinterpreting results that have been interpreted once already, gradually developing a theory—a theory of why flight strips help controllers know what is going on that is anchored in the researcher's continually evolving understanding of the informants' work and their world.

Closing the Gap to the Future

The three high-level categories of controller (flight-strip) work tell certifiers that air-traffic controllers have developed strategies for dealing with the communication of complexity to other controllers, for predicting workload and planning future work. Flight strips play a central, but not necessarily exclusive, role. The research account is written up in such a way that the status quo does not get the prerogative: Tools other than flight strips could conceivably help controllers deal with complexity, dynamics, and coordination issues. Complexity and dynamics, as well as coordination, are critical to what makes air-traffic control what it is, including difficult. Whatever certifiers will want to brand as safe to use, they would do well to take into account that controllers use their artifact(s) to help them deal with complexity, to help them anticipate dynamic futures, and to support their coordination with other controllers. This resembles some kind of human factors requirements that could provide a certifier with meaningful input.

CERTIFYING UNDER UNCERTAINTY

One role of human factors is to help developers and certifiers judge whether a technology is safe for future use. But quantitative and qualitative human factors communities both risk taking the authority of their findings for granted and regarding the translation to future, and claims about the future being either safe or unsafe, as essentially nonproblematic. At least the literature (both literatures) are relatively silent on this fundamental issue. Yet neither the legitimacy of findings nor the translation to claims about the future is in fact easily achieved, or should be taken for granted. More work needs to be done to produce findings that make sense for those who have to certify a system as safe to use. Experimental human factors research can claim empirical legitimacy by virtue of the authority vested in the laboratory researcher and the control over the method used to get data. Such research can speak meaningfully to future use because it tests microversions of a future system. Researchers, however, should explicitly indicate where the versions of the future they tested are impoverished, and what subtle effects of context on their experimental settings could produce findings that diverge from what future users will encounter.

Qualitative research in human factors can claim legitimacy, and relevance to those who need to certify the next system, because of its authentic encounters with the field where people actually carry out the work. Validation emerges from the literature (what others have said about the same and similar contexts) and from interpretation (how theory and evidence make sense of this particular context). Such research can speak meaningfully to certification issues because it allows users to express their preferences, choices and apprehensions. Qualitative human factors research, however, must not stop at recording and replaying informant statements. It must deconfound informant understandings with understandings informed by concepts, theory, analysis and literature.

Human factors work, of whatever kind, can help bridge the gap from research findings to future systems. Research accounts need to be both convincing as science and cast in a language that allows a certifier to look ahead to the future: looking ahead to work and a co-evolution of people and technology in a system that does not yet exist.

Chapter 10

Should We Hold People Accountable for Their Mistakes?

Transportation human factors has made enormous progress over the past decades. It would be easy to claim that transportation systems have become safer in part through human factors efforts. As a result of such work over the past decades, progress on safety has become synonymous with:

- Taking a systems perspective: Accidents are not caused by failures of individuals, but emerge from the conflux or alignment of multiple contributory system factors, each necessary and only jointly sufficient. The source of accidents is the system, not its component parts.
- Moving beyond blame: Blame focuses on the supposed defects of individual operators and denies the import of systemic contributions. In addition, blame has all kinds of negative side effects. It typically leads to defensive posturing, obfuscation of information, protectionism, polarization, and mute reporting systems.

Progress on safety coincides with learning from failure. This makes punishment and learning two mutually exclusive activities: Organizations can either learn from an accident or punish the individuals involved in it, but hardly do both at the same time. The reason is that punishment of individuals can protect false beliefs about basically safe systems, where humans are the least reliable components. Learning challenges and potentially changes the belief about what creates safety. Moreover, punishment emphasizes that failures are deviant, that they do not naturally belong in the organization. Learning means that failures are seen as normal, as resulting from the in-

193

herent pursuit of success in resource-constrained, uncertain environments. Punishment turns the culprits into unique and necessary ingredients for the failure to happen. Punishment, rather than helping people avoid or better manage conditions that are conducive to error, actually conditions people not to get caught when errors do occur. This stifles learning. Finally, punishment is about the search for closure, about moving beyond and away from the adverse event. Learning is about continuous improvement, about closely integrating the event in what the system knows about itself.

Making these ideas stick, however, is not proving as easy as it was to develop them. In the aftermath of several recent accidents and incidents, the operators involved (pilots or air-traffic controllers in these cases) were charged with criminal offenses (e.g., professional negligence, manslaughter). In some accidents even organizational management has been held criminally liable. Criminal charges differ from civil lawsuits in many respects. Most obviously, the target is not an organization, but individuals (air-traffic controllers, flight crew, maintenance technicians). Punishment consists of possible incarceration or some putatively rehabilitative alternative—not (just) financial compensation. Unlike organizations covered against civil suits, few operators or managers themselves have insurance to pay for legal defense against criminal charges that arise from doing their jobs.

Some maintain that criminally pursuing operators or managers for erring on the job is morally unproblematic. The greater good befalls the greater number of people (i.e., all potential passengers) by protecting them from unreliable operators. A lot of people win, only a few outcasts lose. To human factors, however, this may be utilitarianism inverted. Everybody loses when human error gets criminalized: Upon the threat of criminal charges, operators stop sending in safety-related information; incident reporting grinds to a halt. Criminal charges against individual operators also polarize industrial relations. If the organization wants to limit civil liability, then official blame on the operator could deflect attention from upstream organizational issues related to training, management, supervision, and design decisions. Blaming such organizational issues, in contrast, can be a powerful ingredient in an individual operator's criminal defense—certainly when the organization has already rendered the operator expendable by euphemism (standby, ground duty, administrative leave) and without legitimate hope of meaningful re-employment. In both cases, industrial relations are destabilized. Intra-organizational battles become even more complex when individual managers get criminally pursued; defensive maneuvering by these managers typically aims to off-load the burden of blame onto other departments or parts of the organization. This easily leads to poisonous relations and a crippling of organizational functioning. Finally, incarceration or alternative punishment of operators or managers has no demonstrable rehabilitative effect (perhaps because there is nothing to re-habilitate). It does not make an operator or manager any safer, nor is there evidence of vicarious learning (learning by example and fear of punish-ment). Instead, punishment or its threat merely leads to counterproductive responses, to people ducking the debris.

The transportation industry itself shows ambiguity with regard to the criminalization of error. Responding to the 1996 ValuJet accident, where mechanics loaded oxygen generators into the cargo hold of a DC-9, which subsequently caught fire, the editor of *Aviation Week and Space Technology* "strongly believed the failure of SabreTech employees to put caps on oxy-gen generators constituted willful negligence that led to the killing of 110 passengers and crew. Prosecutors were right to bring charges. There has to be some fear that not doing one's job correctly could lead to prosecution" (North, 2000, p. 66). Rescinding this 2 years later, however, North (2002) opined that learning from accidents and criminal prosecution go together like "oil and water, cats and dogs," that "criminal probes do not mix well with aviation accident inquiries" (p. 70). Most other cases reveal similar in-stability with regard to prosecuting operators for error. Culpability in avia-tion does not appear to be a fixed notion, connected unequivocally to fea-tures of some incident or accident. Rather, culpability is a highly flexible category. Culpability is negotiable, subject to national and professional in-terpretations, influenced by political imperatives and organizational pres-sures, and part of personal or institutional histories.

As psychologists point out, culpability is also about assumptions we make about the amount of control people had when carrying out their (now) controversial acts. The problem here is that hindsight deeply confounds such judgments of control. In hindsight, it may seem obvious that people had all the necessary data available to them (and thus the potential for con-trol and safe outcomes). Yet they may have willingly ignored this data in or-der to get home faster, or because they were complacent. Retrospect and the knowledge of outcome deeply affect our ability to judge human per-formance, and a reliance on folk models of phenomena like complacency, situation awareness, and stress does not help. All too quickly we come to the conclusion that people could have better controlled the outcome of a situa-tion, if only they had invested a little more effort.

ACCOUNTABILITY

What is accountability, and what does it actually mean to hold people ac-countable for their mistakes? Social cognition research shows that account-ability or holding people accountable is not that simple. Accountability is fundamental to any social relation. There is always an implicit or explicit expectation that we may be called on to justify our beliefs and actions to others. The social-functionalist argument for accountability is that this expectation is mutual: As social beings we are locked into reciprocating relationships. Accountability, however, is not a unitary concept—even if this is what many stakeholders may think when aiming to improve people's performance under the banner of holding them accountable. There are as many types of accountability as there are distinct relationships among people, and between people and organizations, and only highly specialized subtypes of accountability actually compel people to expend more cognitive effort. Expending greater effort, moreover, does not necessarily mean better task performance, as operators may become concerned more with limiting exposure and liability than with performing well (Lerner & Tetlock, 1999), something that can be observed in the decline of incident reporting with threats of prosecution (North, 2002). What is more, if accounting is perceived as illegitimate, for example, intrusive, insulting, or ignorant of real work, then any beneficial effects of accountability will vanish or backfire. Effects that have been experimentally demonstrated include a decline in motivation, excessive stress, and attitude polarization, and the same effects can be seen in recent cases where pilots and air-traffic controllers were held accountable by courts and other parties ignorant of the real trade-offs and dilemmas that make up actual operational work.

The research base on social cognition, then, tells us that accountability, even if inherent in human relationships, is not unambiguous or unproblematic. The good side of this is that, if accountability can take many forms, then alternative, perhaps more productive avenues of holding people accountable are possible. Giving an account, after all, does not have to mean exposing oneself to liability, but rather, telling one's story so that others can learn vicariously. Many sources, even within human factors, point to the value of storytelling in preparing operators for complex, dynamic situations in which not everything can be anticipated. Stories are easily remembered, scenario-based plots with actors, intentions, clues, and outcomes that in one way or another can be mapped onto current difficult situations and matched for possible ways out. Incident-reporting systems can capitalize on this possibility, whereas more incriminating forms of accountability actually retard this very quality by robbing from people the incentive to tell stories in the first place.

ANTHROPOLOGICAL UNDERSTANDINGS OF BLAME

The anthropologist is not intrigued by flaws in people's reasoning process that produce for example, the hindsight bias, but wants to know something about casting blame. Why is blame a meaningful response for those doing the blaming? Why do we turn error into crime? Mary Douglas (1992) described how peoples are organized in part by the way in which they explain misfortune and subsequently pursue retribution or dispense justice. Societies tend to rely on one dominant model of possible cause from which they construct a plausible explanation. In the moralistic model, for example, misfortune is seen as the result of offending ancestors, of sinning, or of breaking some taboo. The inflated, exaggerated role that procedure violations (one type of sinning or taboo breaking) are given in retrospective accounts of failure represent one such use for moralistic models of breakdown and blame. The moralistic explanation (you broke the rule, then you had an accident) is followed by a fixed repertoire of obligatory actions that follow on that choice. If taboos have been broken, then rehabilitation can be demanded through expiatory actions. Garnering forgiveness through some purification ritual is one example. Forcing operators to publicly offer their apologies is a purification ritual seen in the wake of some accidents. Moreover, the rest of the community is reminded to not sin, to not break the taboos, lest the same fate befall them. How many reminders are there in the transportation industry imploring operators to "follow the rules," to "follow the procedures"? These are moralistic appeals with little demonstrable effect on practice, but they may make industry participants feel better about their systems; they may make them feel more in control.

In the extrogenous model, external enemies of the system are to blame for misfortune, a response that can be observed even today in the demotion or exile of failed operators: pilots or controllers or technicians. These people are *ex post facto* relegated to a kind of underclass that no longer represents the professional corps. Firing them is one option, and is used relatively often. But there are more subtle expressions of the extrogenous model too. The ritualistic expropriation of badges, certificates, stripes, licenses, uniforms, or other identity and status markings in the wake of an accident delegitimizes the errant operator as a member of the operational community. A part of such derogation, of course, is psychological defense on the part of (former) colleagues who would need to distance themselves from a realization of equal vulnerability to similar failures. Yet such delegitimization also makes criminalization easier by beginning the incremental process of dehumanizing the operator in question. Wilkinson (1994) presented an excellent example of such demonizing in the consequences that befell a Boeing 747 pilot after allegedly narrowly missing a hotel at London Heathrow airport in thick fog. Demonizing there was incremental in the sense that it made criminal pursuit not only possible in the first place, but subsequently necessary. It fed on itself: Demons such as this pilot would need to be punished, demoted, exorcised. The press had a large share in dramatizing the case, promoting the captain's dehumanization to the point where his suicide was the only way out.

Failure and Fear

Today, almost every misfortune is followed by questions centering on "whose fault?" and "what damages, compensation?" Every death must be chargeable to somebody's account. Such responses approximate the primitives' resistance to the idea of natural death remarkably well (Douglas, 1992). Death, even today, is not considered natural—it has to arise from some type of identifiable cause. Such resistance to the notion that deaths actually can be accidental is obvious in responses to recent mishaps. For example, Snook (2000) commented on his own disbelief, his struggle, in analyzing the friendly shoot-down of two U.S. Black Hawk helicopters by U.S. Fighter Jets over Northern Iraq in 1993:

> This journey played with my emotions. When I first examined the data, I went in puzzled, angry, and disappointed—puzzled how two highly trained Air Force pilots could make such a deadly mistake; angry at how an entire crew of AWACS controllers could sit by and watch a tragedy develop without taking action; and disappointed at how dysfunctional Task Force OPC must have been to have not better integrated helicopters into its air operations. Each time I went in hot and suspicious. Each time I came out sympathetic and unnerved. . . . If no one did anything wrong; if there were no unexplainable surprises at any level of analysis; if nothing was abnormal from a behavioral and organizational perspective; then what *have* we learned? (p. 203)

Snook (2000) confronted the question of whether learning, or any kind of progress on safety, is possible at all if we can find no wrongdoing, no surprises, if we cannot find some kind of deviance. If everything was normal, then how could the system fail? Indeed, this must be among the greater fears that define Western society today. Investigations that do not turn up a "Eureka part," as the label became in the TWA800 probe, are feared not because they are bad investigations, but because they are scary. Philosophers like Nietzsche pointed out that the need for finding a cause is fundamental to human nature. Not being able to find a cause is profoundly distressing; it creates anxiety because it implies a loss of control. The desire to find a cause is driven by fear. So what do we do if there is no Eureka part, no fault nucleus, no seed of destruction? Is it possible to acknowledge that failure results from normal people doing business as usual in normal organizations? Not even many accident investigations succeed at this. As Galison (2000) noted:

> If there is no seed, if the bramble of cause, agency, and procedure does not issue from a fault nucleus, but is rather unstably perched between scales, between human and non-human, and between protocol and judgment, then the world is a more disordered and dangerous place. Accident reports, and

much of the history we write, struggle, incompletely and unstably, to hold that nightmare at bay. (p. 32)

Galison's (2000) remarks remind us of this fear (this nightmare) of not being in control over the systems we design, build, and operate. We dread the possibility that failures emerge from the intertwined complexity of normal everyday systems interactions. We would rather see failures emanate from a traceable, controllable single seed or nucleus. In assigning cause, or in identifying our imagined core of failure, accuracy does not seem to matter. Being afraid is worse than being wrong. Selecting a scapegoat to carry the interpretive load of an accident or incident is the easy price we pay for our illusion that we actually have control over our risky technologies. This price is the inevitable side effect of the centuries-old pursuit of Baconian control and technological domination over nature. Sending controllers, or pilots, or maintenance technicians to jail may be morally wrenching (but not unequivocally so—remember North, 2000), but it is preferable over its scary alternative: acknowledging that we do not enjoy control over the risky technologies we build and consume. The alternative would force us to really admit that failure is an emergent property, that "mistake, mishap and disaster are socially organized and systematically produced by social structures," that these mistakes are normal, to be expected because they are "embedded in the banality of organizational life" (Vaughan, 1996, p. xiv). It would force us to acknowledge the relentless inevitability of mistake in organizations, to see that harmful outcomes can occur in the organizations constructed to prevent them, that harmful consequences can occur even when everybody follows the rules.

Preferring to be wrong over being afraid in the identification of cause overlaps with the common reflex toward individual responsibility in the West. Various transportation modes (particularly aviation) have exported this bias to less individually oriented cultures as well. In the Western intellectual tradition since the Scientific Revolution, it has seemed self-evident to evaluate ourselves as individuals, bordered by the limits of our minds and bodies, and evaluated in terms of our own personal achievements. From the Renaissance onward, the individual became a central focus, fueled in part by Descartes' psychology that created "self-contained individuals" (Heft, 2001). The rugged individualism developed on the back of mass European immigration into North America in the late 19th and early 20th centuries accelerated the image of independent, free heroes accomplishing greatness against all odds, and antiheroes responsible for disproportionate evildoing (e.g., Al Capone). Lone antiheroes still play the lead roles in our stories of failure. The notion that it takes teamwork, or an entire organization, an entire industry (think about Alaska 261) to break a system is just too eccentric relative to this cultural prejudice.

There are earlier bases for the dominance of individualism in Western traditions as well. Saint Augustine, the deeply influential moral thinker for Judeo-Christian societies, saw human suffering as occurring not only because of individual human fault (Pagels, 1988), but because of human choice, the conscious, deliberate, rational choice to err. The idea of a rational choice to err is so pervasive in Western thinking that it goes virtually unnoticed, unquestioned, because it makes such common sense. The idea, for example, is that pilots have a choice to take the correct runway but fail to take it. Instead, they make the wrong choice because of attentional deficiencies or motivational shortcomings, despite the cues that were available and the time they had to evaluate those cues. Air-traffic controllers have a choice to see a looming conflict, but elect to pay no attention to it because they think their priorities should be elsewhere. After the fact, it often seems as if people chose to err, despite all available evidence indicating they had it wrong.

The story of Adam's original sin, and especially what Saint Augustine made of it, reveals the same space for conscious negotiation that we retrospectively invoke on behalf of people carrying out safety-critical work in real conditions. Eve had a deliberative conversation with the snake on whether to sin or not to sin, on whether to err or not to err. The allegory emphasizes the same conscious presence of cues and incentives to not err, of warnings to follow rules and not sin, and yet Adam and Eve elected to err anyway. The prototypical story of error and violation and its consequences in Judeo-Christian tradition tells of people who were equipped with the requisite intellect, who had received the appropriate indoctrination (don't eat that fruit), who displayed capacity for reflective judgment, and who actually had the time to choose between a right and a wrong alternative. They then proceeded to pick the wrong alternative, a choice that would make a big difference for their lives and the lives of others. It is likely that, rather than causing the fall into continued error, as Saint Augustine would have it, Adam's original sin portrays how we think about error, and how we have thought about it for ages. The idea of free will permeates our moral thinking, and most probably influences how we look at human performance to this day.

MISMATCH BETWEEN AUTHORITY AND RESPONSIBILITY

Of course this illusion of free will, though dominant in post hoc analyses of error, is at odds with the real conditions under which people perform work: where resource limitations and uncertainty severely constrain the choices open to them. Van den Hoven (2001) called this "the pressure condition." Operators such as pilots and air-traffic controllers are "narrowly embedded"; they are "configured in an environment and assigned a place which will provide them with observational or derived knowledge of relevant facts and states of affairs" (p. 3). Such environments are exceedingly hostile to the kind of reflection necessary to meet the regulative ideal of individual moral responsibility. Yet this is exactly the kind of reflective idyll we read in the story of Adam and Eve and the kind we retrospectively presume on behalf of operators in difficult situations that led to a mishap.

Human factors refers to this as an authority–responsibility double bind: A mismatch occurs between the responsibility expected of people to do the right thing, and the authority given or available to them to live up to that responsibility. Society expresses its confidence in operators' responsibility through payments, status, symbols, and the like. Yet operators' authority may fall short of that responsibility in many important ways. Operators typically do not have the degrees of freedom assumed by their professional responsibility because of a variety of reasons: Practice is driven by multiple goals that may be incompatible (simultaneously having to achieve maximum capacity utilization, economic aims, customer service, and safety). As Wilkinson (1994, p. 87) remarked: "A lot of lip service is paid to the myth of command residing in the cockpit, to the fantasy of the captain as ultimate decision-maker. But today the commander must first consult with the accountant." Error, then, must be understood as the result of constraints that the world imposes on people's goal-directed behavior. As the local rationality principle dictates, people want to do the right thing, yet features of their work environment limit their authority to act, limit their ability to live up to the responsibility for doing the right thing. This moved Claus Jensen (1996) to say:

> there is no longer any point in appealing to the individual worker's own sense of responsibility, morality or decency, when almost all of us are working within extremely large and complex systems . . . According to this perspective, there is no point in expecting or demanding individual engineers or managers to be moral heroes; far better to put all of one's efforts into reinforcing safety procedures and creating structures and processes conducive to ethical behavior. (p. xiii)

Individual authority, in other words, is constrained to the point where moral appeals to individual responsibility are becoming useless. And authority is not only restricted because of the larger structures that people are only small parts of. Authority to assess, decide, and act can be in limited simply because of the nature of the situation. Time and other resources for making sense of a situation are lacking; information may not be at hand or may be ambiguous; there may be all kinds of subtle organizational pressures to prefer certain actions over others; and there may be no neutral or additional expertise to draw on. Even Eve was initially alone with the snake.

Where was Adam, the only other human available in paradise during those critical moments of seduction into error? Only recent additions to the human factors literature (e.g., naturalistic decision making, ecological task analyses) explicitly took these and other constraints on people's practice into consideration in the design and understanding of work. Free will is a logical impossibility in cases where there is a mismatch between responsibility and authority, which is to say that free will is always a logical impossibility in real settings where real safety-critical work is carried out.

This should invert the culpability criterion when operators or others are being held accountable for their errors. Today it is typically the defendant who has to explain that he or she was constrained in ways that did not allow adequate control over the situation. But such defenses are often hopeless. Outsider observers are influenced by hindsight when they look back on available data and choice moments. As a consequence, they consistently overestimate both the clarity of the situation and the ability to control the outcome. So rather than the defendant having to show that insufficient data and control made the outcome inevitable, it should be up to the claimants, or prosecution, to prove that adequate control was in fact available. Did people have enough authority to live up to their responsibility?

Such a proposal, however, amounts to only a marginal adjustment of what may still be dysfunctional and counterproductive accountability relationships. What different models of responsibility could possibly replace current accountability relationships, and do they have any chance? In the adversarial confrontations and defensive posturing that the criminalization of error generates today, truth becomes fragmented across multiple versions that advocate particular agendas (staying out of jail, limiting corporate liability). This makes learning from the mishap almost impossible. Even making safety improvements in the wake of an accident can get construed as an admission of liability. This robs systems of their most concrete demonstration that they have learned something from the mishap: an actual implementation of lessons learned. Indeed, lessons are not learned before organizations have actually made the changes that those lessons prescribe.

BLAME-FREE CULTURES?

Ideally, there should be accountability without invoking defense mechanisms. Blame-free cultures, for example, though free from blame and associated protective plotting, are not without member responsibility. But blame-free cultures are extremely rare. Examples have been found among Sherpas in Nepal (Douglas, 1992), who pressure each other to settle quarrels peacefully and reduce rivalries with strong informal procedures for reconciliation. Laying blame accurately is considered much less important than a generous treatment of the victim. Sherpas irrigate their social system with a lavish flow of gifts, taxing themselves collectively to ensure nobody goes neglected, and victims are not left exposed to impoverishment or discrimination (Douglas). This mirrors the propensity of Scandinavian cultures for collective taxation to support dense webs of social security. Prosecution of individuals or especially civil lawsuits in the wake of accidents are rare. U.S. responses stand in stark contrast (although criminal prosecution of operators is rare there). Despite a plenitude of litigation (which inflates and occasionally exceeds the compensatory expectations of a few), victims as a group are typically undercompensated. Blame-free cultures may hinge more on consistently generous treatment of victims than on denying that professional accountability exists. They also hinge on finding other expressions of responsibility, of what it means to be a responsible member of that culture.

Holding people accountable can be consistent with being blame-free if transportation industries think in novel ways about accountability. This would involve innovations in relationships among the various stakeholders. Indeed, in order to continue making progress on safety, transportation industries should reconsider and reconstruct accountability relationships between its stakeholders (organizations, regulators, litigators, operators, passengers). In a new form of accountability relationships, operators or managers involved in mishaps could be held accountable by inviting them to tell their story (their account). Such accounts can then be systematized and distributed, and used to propagate vicarious learning for all. Microversions of such accountability relationships have been implemented in many incident-reporting systems, and perhaps their examples could move industries in the direction of as yet elusive blame-free cultures.

The odds, however, may be stacked against attempts to make such progress. The Judeo-Christian ethic of individual responsibility is not just animated by a basic Nietzschean anxiety of losing control. Macrostructural forces are probably at work too. There is evidence that episodes of renewed enlightenment, such as the Scientific Revolution, are accompanied by violent regressions toward supernaturalism and witch hunting. Prima facie, this would be an inconsistency. How can an increasingly illuminated society simultaneously retard into superstition and scapegoating? One answer may lie in the uncertainties and anxieties brought on by the technological advances and depersonalization that inevitably seem to come with such progress. New, large, complex, and widely extended technological systems (e.g., global aviation that took just a few decades to expand into what it is today) create displacement, diffusion, and causal uncertainty. A reliance on individual culpability may be the only sure way of recapturing an illusion of control. In contrast, less technologically or industrially developed societies (take the Sherpas as example again) appear to rely on more benign models of failure and blame, and more on collective responsibility.

In addition, those who do safety-critical work often tie culpability conventions to aspects of their personal biographies. Physician Atul Gawande (2002, p. 73), for example, commented on a recent surgical incident and observed that terms such as *systems problems* are part of a "dry language of structures, not people . . . something in me, too, demands an acknowledgement of my autonomy, which is also to say my ultimate culpability . . . although the odds were against me, it wasn't as if I had no chance of succeeding. Good doctoring is all about making the most of the hand you're dealt, and I failed to do so."

The expectation of being held accountable if things go wrong (and, conversely, being responsible if things go right) appears intricately connected to issues of self-identity, where accountability is the other side of professional autonomy and a desire for control. This expectation can engender considerable pride and can make even routine operational work deeply meaningful. But although good doctoring (or any kind of practice) may be making the most of the hand one is dealt, human factors has always been about providing that hand more and better opportunities to do the right thing. Merely leaving the hand with what it is dealt and banking on personal motivation to do the rest takes us back to prehistoric times, when behaviorism reigned and human factors had yet to make its entry in system safety thinking.

Accountability and culpability are deeply complex concepts. Disentangling their prerational influences in order to promote systems thinking, and to create an objectively fairer, blame-free culture, may be an uphill struggle. They are, in any case, topics worthy of more research.

References

Aeronautica Civil. (1996). *Aircraft accident report: Controlled flight into terrain, American Airlines flight 965, Boeing 757-223, N651AA near Cali, Colombia, December 20, 1995.* Bogota, Colombia: Author.

Airliner World. (2001, November). *Excel,* pp. 77–80.

Air Transport Association of America. (1989, April). *National plan to enhance aviation safety through human factors improvements.* Washington, DC: Author.

Albright, C. A., Truitt, T. R., Barile, A. B., Vortac, O. U., & Manning, C. A. (1996). *How controllers compensate for the lack of flight progress strips* (Final Rep. No. DOT/FAA/AM-96/5). Arlington, VA: National Technical Information Service.

Amalberti, R. (2001). The paradoxes of almost totally safe transportation systems. *Safety Science, 37,* 109–126.

Angell, I. O., & Straub, B. (1999). Rain-dancing with pseudo-science. *Cognition, Technology and Work, 1,* 179–196.

Baiada, R. M. (1995). ATC biggest drag on airline productivity. *Aviation Week and Space Technology, 31,* 51–53.

Bainbridge, L. (1987). Ironies of automation. In J. Rasmussen, K. Duncan, & J. Leplat (Eds.), *New technology and human error* (pp. 271–283). Chichester, England: Wiley.

Batteau, A. W. (2001). The anthropology of aviation and flight safety. *Human Organization, 60*(3), 201–210.

Beyer, H., & Holtzblatt, K. (1998). *Contextual design: Defining customer-centered systems.* San Diego, CA: Academic Press.

Billings, C. E. (1996). Situation awareness measurement and analysis: A commentary. In D. J. Garland & M. R. Endsley (Eds.), *Experimental analysis and measurement of situation awareness* (pp. 1–5). Daytona Beach, FL: Embry-Riddle Aeronautical University Press.

Billings, C. E. (1997). *Aviation automation: The search for a human-centered approach.* Mahwah, NJ: Lawrence Erlbaum Associates.

Björklund, C., Alfredsson, J., & Dekker, S. W. A. (2003). Shared mode awareness of the FMA in commercial aviation: An eye-point of gaze and communication data analysis in a high-fidelity simulator. In E. Hollnagel (Ed.), *Proceedings of EAM 2003, The 22nd European Confer-*

ence on Human Decision Making and Manual Control (pp. 119–126). Linköping, Sweden: Cognitive Systems Engineering Laboratory, Linköping University.

Boeing Commercial Airplane Group. (1996). *Boeing submission to the American Airlines Flight 965 Accident Investigation Board.* Seattle, WA: Author.

Bruner, J. (1990). *Acts of meaning.* Cambridge, MA: Harvard University Press.

Campbell, R. D., & Bagshaw, M. (1991). *Human performance and limitations in aviation.* Oxford, England: Blackwell Science.

Capra, F. (1982). *The turning point.* New York: Simon & Schuster.

Carley, W. M. (1999, January 21). Swissair pilots differed on how to avoid crash. *The Wall Street Journal.*

Columbia Accident Investigation Board. (2003). Report Volume 1, August 2003. Washington, DC: U.S. Government Printing Office.

Cordesman, A. H., & Wagner, A. R. (1996). *The lessons of modern war: Vol. 4. The Gulf War.* Boulder, CO: Westview Press.

Croft, J. (2001, July 16). Researchers perfect new ways to monitor pilot performance. *Aviation Week and Space Technology*, pp. 76–77.

Dawkins, R. (1986). *The blind watchmaker.* London: Penguin.

Degani, A., Heymann, M., & Shafto, M. (1999). Formal aspects of procedures: The problem of sequential correctness. In *Proceedings of the 43rd Annual Meeting of the Human Factors and Ergonomics Society.* Houston, TX: Human Factors Society.

Dekker, S. W. A. (2002). *The field guide to human error investigations.* Bedford, England: Cranfield University Press.

Dekker, S. W. A., & Woods, D. D. (1999). To intervene or not to intervene: The dilemma of management by exception. *Cognition, Technology and Work, 1*, 86–96.

Della Rocco, P. S., Manning, C. A., & Wing, H. (1990). *Selection of air traffic controllers for automated systems: Applications from current research* (DOT/FAA/AM-90/13). Arlington, VA: National Technical Information Service.

Dörner, D. (1989). *The logic of failure: Recognizing and avoiding error in complex situations.* Cambridge, MA: Perseus Books.

Douglas, M. (1992). *Risk and blame: Essays in cultural theory.* London: Routledge.

Endsley, M. R., Mogford, M., Allendoerfer, K., & Stein, E. (1997). *Effect of free flight conditions on controller performance, workload and situation awareness: A preliminary investigation of changes in locus of control using existing technologies.* Lubbock, TX: Texas Technical University.

Feyerabend, P. (1993). *Against method* (3rd ed.). London: Verso.

Feynman, R. P. (1988). *"What do you care what other people think?": Further adventures of a curious character.* New York: Norton.

Fischoff, B. (1975). Hindsight is not foresight: The effect of outcome knowledge on judgement under uncertainty. *Journal of Experimental Psychology: Human Perception and Performance, 1*(3), 288–299.

Fitts, P. M. (1951). *Human engineering for an effective air navigation and traffic control system.* Washington, DC: National Research Council.

Fitts, P. M., & Jones, R. E. (1947). *Analysis of factors contributing to 460 "pilot error" experiences in operating aircraft controls* (Memorandum Rep. No. TSEAA-694-12). Dayton, OH: Aero Medical Laboratory, Air Material Command, Wright-Patterson Air Force Base.

Flores, F., Graves, M., Hartfield, B., & Winograd, T. (1988). Computer systems and the design of organizational interaction. *ACM Transactions on Office Information Systems, 6*, 153–172.

Galison, P. (2000). An accident of history. In P. Galison & A. Roland (Eds.), *Atmospheric flight in the twentieth century* (pp. 3–44). Dordrecht, The Netherlands: Kluwer Academic.

Galster, S. M., Duley, J. A., Masolanis, A. J., & Parasuraman, R. (1999). Effects of aircraft self-separation on controller conflict detection and workload in mature Free Flight. In M. W.

Scerbo & M. Mouloua (Eds.), *Automation technology and human performance: Current research and trends* (pp. 96–101). Mahwah, NJ: Lawrence Erlbaum Associates.

Gawande, A. (2002). *Complications: A surgeon's notes on an imperfect science.* New York: Picado.

Geertz, C. (1973). *The interpretation of cultures.* New York: Basic Books.

Golden-Biddle, K., & Locke, K. (1993). Appealing work: An investigation of how ethnographic texts convince. *Organization Science, 4,* 595–616.

Heft, H. (2001). *Ecological psychology in context: James Gibson, Roger Barker, and the legacy of William James's radical empiricism.* Mahwah, NJ: Lawrence Erlbaum Associates.

Helmreich, R. L. (2000). On error management: Lessons from aviation. *British Medical Journal, 320,* 745–753.

Helmreich, R. L., Klinect, J. R., & Wilhelm, J. A. (1999). Models of threat, error and response in flight operations. In R. S. Jensen (Ed.), *Proceedings of the 10th International Symposium on Aviation Psychology.* Columbus: The Ohio State University.

Hollan, J., Hutchins, E., & Kirsh, D. (2000). Distributed cognition: Toward a new foundation for human–computer interaction research. *ACM Transactions on Computer–Human Interaction, 7*(2), 174–196.

Hollnagel, E. (1999). From function allocation to function congruence. In S. W. A. Dekker & E. Hollnagel (Eds.), *Coping with computers in the cockpit* (pp. 29–53). Aldershot, England: Ashgate.

Hollnagel, E. (Ed.). (2003). *Handbook of cognitive task design.* Mahwah, NJ: Lawrence Erlbaum Associates.

Hollnagel, E., & Amalberti, R. (2001). The emperor's new clothes: Or whatever happened to "human error"? In S. W. A. Dekker (Ed.), *Proceedings of the 4th International Workshop on Human Error, Safety and Systems Development* (pp. 1–18). Linköping, Sweden: Linköping University.

Hollnagel, E., & Woods, D. D. (1983). Cognitive systems engineering: New wine in new bottles. *International Journal of Man-Machine Studies, 18,* 583–600.

Hughes, J. A., Randall, D., & Shapiro, D. (1993). From ethnographic record to system design: Some experiences from the field. *Computer Supported Collaborative Work, 1,* 123–141.

International Civil Aviation Organization. (1998). *Human factors training manual* (ICAO Doc. No. 9683-AN/950). Montreal, Quebec: Author.

Jensen, C. (1996). *No downlink: A dramatic narrative about the Challenger accident and our time.* New York: Farrar, Strauss, Giroux.

Joint Aviation Authorities. (2001). *Human factors in maintenance working group report.* Hoofddorp, The Netherlands: Author.

Joint Aviation Authorities. (2003). *Advisory Circular Joint ACJ 25.1329: Flight guidance system, Attachment 1 to NPA (Notice of Proposed Amendment) 25F-344.* Hoofddorp, The Netherlands: Author.

Kern, T. (1998). *Flight discipline.* New York: McGraw-Hill.

Klein, G. A. (1998). *Sources of power: How people make decisions.* Cambridge, MA: MIT Press.

Kohn, L. T., Corrigan, J. M., & Donaldson, M. (Eds.). (1999). *To err is human: Building a safer health system.* Washington, DC: Institute of Medicine.

Kuhn, T. S. (1962). *The structure of scientific revolutions.* Chicago: University of Chicago Press.

Langewiesche, W. (1998). *Inside the sky: A meditation on flight.* New York: Pantheon Books.

Lanir, Z. (1986). *Fundamental surprise.* Eugene, OR: Decision Research.

Lautman, L., & Gallimore, P. L. (1987). Control of the crew caused accident: Results of a 12-operator survey. *Boeing Airliner,* April–June, 1–6.

Lerner, J. S., & Tetlock, P. E. (1999). Accounting for the effects of accountability. *Psychological Bulletin, 125,* 255–275.

Leveson, N. (2002). *A new approach to system safety engineering.* Cambridge, MA: Aeronautics and Astronautics, Massachusetts Institute of Technology.

Mackay, W. E. (2000). Is paper safer? The role of paper flight strips in air traffic control. *ACM/ Transactions on Computer-Human Interactions, 6,* 311–340.

McDonald, N., Corrigan, S., & Ward, M. (2002, June). *Well-intentioned people in dysfunctional systems.* Keynote paper presented at the 5th Workshop on Human Error, Safety and Systems Development, Newcastle, Australia.

Meister, D. (2003). The editor's comments. *Human Factors Ergonomics Society COTG Digest, 5,* 2–6.

Metzger, U., & Parasuraman, R. (1999). Free Flight and the air traffic controller: Active control versus passive monitoring. In *Proceedings of the Human Factors and Ergonomics Society 43rd annual meeting.* Houston, TX: Human Factors Society.

Mumaw, R. J., Sarter, N. B., & Wickens, C. D. (2001). Analysis of pilots' monitoring and performance on an automated flight deck. In *Proceedings of 11th International Symposium in Aviation Psychology.* Columbus: Ohio State University.

National Aeronautics and Space Administration. (2000, March). *Report on project management in NASA, by the Mars Climate Orbiter Mishap Investigation Board.* Washington, DC: Author.

National Transportation Safety Board. (1974). *Delta Air Lines Douglas DC-9-31, Boston, MA, 7/31/73* (NTSB Rep. No. AAR-74/03). Washington, DC: Author.

National Transportation Safety Board. (1995). *Aircraft accident report: Flight into terrain during missed approach, USAir flight 1016, DC-9-31, N954VJ, Charlotte Douglas International Airport, Charlotte, North Carolina, July 2, 1994* (NTSB Rep. No. AAR-95/03). Washington, DC: Author.

National Transportation Safety Board. (1997). *Grounding of the Panamanian passenger ship* Royal Majesty *on Rose and Crown shoal near Nantucket, Massachusetts, June 10, 1995* (NTSB Rep. No. MAR-97/01). Washington, DC: Author.

National Transportation Safety Board. (2002). *Loss of control and impact with Pacific Ocean, Alaska Airlines Flight 261 McDonnell Douglas MD-83, N963AS, about 2.7 miles north of Anacapa Island, California, January 31, 2000* (NTSB Rep. No. AAR-02/01). Washington, DC: Author.

Neisser, U. (1976). *Cognition and reality: Principles and implications of cognitive psychology.* San Francisco: Freeman Press.

North, D. M. (2000, May 15). Let judicial system run its course in crash cases. *Aviation Week and Space Technology,* p. 66.

North, D. M. (2002, February 4). Oil and water, cats and dogs. *Aviation Week and Space Technology,* p. 70.

O'Hare, D., & Roscoe, S. (1990). *Flightdeck performance: The human factor.* Ames: Iowa State University Press.

Orasanu, J. M. (2001). The role of risk assessment in flight safety: Strategies for enhancing pilot decision making. In *Proceedings of the 4th International Workshop on Human Error, Safety and Systems Development* (pp. 83–94). Linköping, Sweden: Linköping University.

Orasanu, J. M., & Connolly, T. (1993). The reinvention of decision making. In G. A. Klein, J. Orasanu, R. Calderwood, & C. E. Zsambok (Eds.), *Decision making in action: Models and methods* (pp. 3–20). Norwood, NJ: Ablex.

Pagels, E. (1988). *Adam, Eve and the serpent.* London: Weidenfeld & Nicolson.

Parasuraman, R., Molly, R., & Singh, I. (1993). Performance consequences of automation-induced complacency. *The International Journal of Aviation Psychology, 3*(1), 1–23.

Parasuraman, R., Sheridan, T. B., & Wickens, C. D. (2000). A model for types and levels of human interaction with automation. IEEE transactions on systems, man, and cybernetics—Part A: Systems and Humans. *Systems and Humans, 30,* 286–297.

Perrow, C. (1984). *Normal accidents: Living with high-risk technologies.* New York: Basic Books.

Rasmussen, J., & Svedung, I. (2000). *Proactive risk management in a dynamic society.* Karlstad, Sweden: Swedish Rescue Services Agency.

Reason, J. T. (1990). *Human error*. Cambridge, England: Cambridge University Press.

Reason, J. T., & Hobbs, A. (2003). *Managing maintenance error: A practical guide*. Aldershot, England: Ashgate.

Rochlin, G. I. (1999). Safe operation as a social construct. *Ergonomics, 42*, 1549–1560.

Rochlin, G. I., LaPorte, T. R., & Roberts, K. H. (1987). The self-designing high-reliability organization: Aircraft carrier flight operations at sea. *Naval War College Review*, Autumn 1987.

Ross, G. (1995). *Flight strip survey report*. Canberra, Australia: TAAATS TOI.

Sacks, O. (1998). *The man who mistook his wife for a hat*. New York: Touchstone.

Sanders, M. S., & McCormick, E. J. (1997). *Human factors in engineering and design* (7th ed.). New York: McGraw-Hill.

Sarter, N. B., & Woods, D. D. (1997). Teamplay with a powerful and independent agent: A corpus of operational experiences and automation surprises on the Airbus A320. *Human Factors, 39*, 553–569.

Shappell, S. A., & Wiegmann, D. A. (2001). Applying reason: The human factors analysis and classification system (HFACS). *Human Factors and Aerospace Safety, 1*, 59–86.

Singer, G., & Dekker, S. W. A. (2000). Pilot performance during multiple failures: An empirical study of different warning systems. *Journal of Transportation Human Factors, 2*, 63–76.

Smith, K. (2001). Incompatible goals, uncertain information and conflicting incentives: The dispatch dilemma. *Human Factors and Aerospace Safety, 1*, 361–380.

Snook, S. A. (2000). *Friendly fire: The accidental shootdown of US Black Hawks over Northern Iraq*. Princeton, NJ: Princeton University Press.

Starbuck, W. H., & Milliken, F. J. (1988). Challenger: Fine-tuning the odds until something breaks. *Journal of Management Studies, 25*, 319–340.

Statens Haverikommision [Swedish Accident Investigation Board]. (2000). *Tillbud vid landning med flygplanet LN-RLF den 23/6 på Växjö/Kronoberg flygplats, G län* (Rapport RL 2000:38) [Incident during landing with aircraft LN-RLF on June 23 at Växjö/Kronoberg airport]. Stockholm, Sweden: Author.

Statens Haverikommision [Swedish Accident Investigation Board]. (2003). *Tillbud mellan flygplanet LN-RPL och en bogsertraktor på Stockholm/Arlanda flygplats, AB län, den 27 oktober 2002* (Rapport RL 2003:47) [Incident between aircraft LN-RPL and a tow-truck at Stockholm/Arlanda airport, October 27, 2002]. Stockholm, Sweden: Author.

Suchman, L. A. (1987). *Plans and situated actions: The problem of human-machine communication*. Cambridge, England: Cambridge University Press.

Tuchman, B. W. (1981). *Practicing history: Selected essays*. New York: Norton.

Turner, B. (1978). *Man-made disasters*. London: Wykeham.

Varela, F. J., Thompson, E., & Rosch, E. (1991). *The embodied mind: Cognitive science and human experience*. Cambridge, MA: MIT Press.

Vaughan, D. (1996). *The Challenger lauch decision: Risky technology, culture and deviance at NASA*. Chicago: University of Chicago Press.

Vaughan, D. (1999). The dark side of organizations: Mistake, misconduct, and disaster. *Annual Review of Sociology, 25*, 271–305.

van den Hoven, M. J. (2001). *Moral responsibility and information technology*. Rotterdam, The Netherlands: Erasmus University Center for Philosophy of ICT.

Vicente, K. (1999). *Cognitive work analysis: Toward safe, productive, and healthy computer-based work*. Mahwah, NJ: Lawrence Erlbaum Associates.

Weick, K. E. (1993). The collapse of sensemaking in organizations. *Administrative Science Quarterly, 38*, 628–652.

Weick, K. E. (1995). *Sensemaking in organizations*. London: Sage.

Weingart, P. (1991). Large technical systems, real life experiments, and the legitimation trap of technology assessment: The contribution of science and technology to constituting risk perception. In T. R. LaPorte (Ed.), *Social responses to large technial systems: Control or anticipation* (pp. 8–9). Amsterdam: Kluwer.

Wiener, E. L. (1988). Cockpit automation. In E. L. Wiener & D. C. Nagel (Eds.), *Human factors in aviation* (pp. 433–462). San Diego, CA: Academic Press.

Wilkinson, S. (1994, February–March). The Oscar November incident. *Air & Space*, 80–87.

Woods, D. D. (1993). Process-tracing methods for the study of cognition outside of the experimental laboratory. In G. A. Klein, J. Orasanu, R. Calderwood, & C. E. Zsambok (Eds.), *Decision making in action: Models and methods* (pp. 228–251). Norwood, NJ: Ablex.

Woods, D. D. (2003, October 29). *Creating foresight: How resilience engineering can transform NASA's approach to risky decision making.* Hearing before the U.S. Senate Committee on Commerce, Science and Transportation, John McCain, chair, Washington, DC.

Woods, D. D., & Dekker, S. W. A. (2001). Anticipating the effects of technology change: A new era of dynamics for Human Factors. *Theoretical Issues in Ergonomics Science, 1,* 272–282.

Woods, D. D., Johannesen, L. J., Cook, R. I., & Sarter, N. B. (1994). *Behind human error: Cognitive systems, computers and hindsight.* Dayton, OH: CSERIAC.

Woods, D. D., Patterson, E. S., & Roth, E. M. (2002). Can we ever escape from data overload? A cognitive systems diagnosis. *Cognition, Technology, and Work, 4,* 22–36.

Woods, D. D., & Shattuck, L. G. (2000). Distant supervision: Local action given the potential for surprise. *Cognition Technology and Work, 2*(4), 242–245.

Wright, P. C., & McCarthy, J. (2003). Analysis of procedure following as concerned work. In E. Hollnagel (Ed.), *Handbook of cognitive task design* (pp. 679–700). Mahwah, NJ: Lawrence Erlbaum Associates.

Wynne, B. (1988). Unruly technology: Practical rules, impractical discourses, and public understanding. *Social Studies of Sciences, 18,* 147–167.

Xiao, Y., & Vicente, K. J. (2000). A framework for epistemological analysis in empirical (laboratory and field) studies. *Human Factors, 42,* 87–101.

Yerkes, R. M., & Dodson, J. D. (1908). The relation of strength of stimulus to rapidity of habit-formation. *Journal of Comparative and Neurological Psychology, 18,* 459–482.

Author Index

Subject Index

Note: Figures in *italics* refer to figures.